CANCER ETIOLOGY, DIAGNOSIS AND TREATMENTS

BRAIN-CANCER ASSOCIATED TUMOR MARKER GENES EXPRESSION PATTERN IN HUMANS

CANCER ETIOLOGY, DIAGNOSIS AND TREATMENTS

Additional books in this series can be found on Nova's website under the Series tab.

Additional E-books in this series can be found on Nova's website under the E-books tab.

CANCER ETIOLOGY, DIAGNOSIS AND TREATMENTS

BRAIN-CANCER ASSOCIATED TUMOR MARKER GENES EXPRESSION PATTERN IN HUMANS

HARUN M. SAID,
CARSTEN HAGEMANN
AND
ADRIAN STAAB

Nova Science Publishers, Inc.
New York

Copyright © 2011 by Nova Science Publishers, Inc.

All rights reserved. No part of this book may be reproduced, stored in a retrieval system or transmitted in any form or by any means: electronic, electrostatic, magnetic, tape, mechanical photocopying, recording or otherwise without the written permission of the Publisher.

For permission to use material from this book please contact us:
Telephone 631-231-7269; Fax 631-231-8175
Web Site: http://www.novapublishers.com

NOTICE TO THE READER

The Publisher has taken reasonable care in the preparation of this book, but makes no expressed or implied warranty of any kind and assumes no responsibility for any errors or omissions. No liability is assumed for incidental or consequential damages in connection with or arising out of information contained in this book. The Publisher shall not be liable for any special, consequential, or exemplary damages resulting, in whole or in part, from the readers' use of, or reliance upon, this material. Any parts of this book based on government reports are so indicated and copyright is claimed for those parts to the extent applicable to compilations of such works.

Independent verification should be sought for any data, advice or recommendations contained in this book. In addition, no responsibility is assumed by the publisher for any injury and/or damage to persons or property arising from any methods, products, instructions, ideas or otherwise contained in this publication.

This publication is designed to provide accurate and authoritative information with regard to the subject matter covered herein. It is sold with the clear understanding that the Publisher is not engaged in rendering legal or any other professional services. If legal or any other expert assistance is required, the services of a competent person should be sought. FROM A DECLARATION OF PARTICIPANTS JOINTLY ADOPTED BY A COMMITTEE OF THE AMERICAN BAR ASSOCIATION AND A COMMITTEE OF PUBLISHERS.

Additional color graphics may be available in the e-book version of this book.

Library of Congress Cataloging-in-Publication Data
Said, Harun M.
Brain-cancer associated tumor marker genes expression pattern in humans / Harun M. Said, Adrian Staab, and Carsten Hagemann.
  p. ; cm.
  Includes bibliographical references and index.
  ISBN 978-1-61728-011-5 (softcover)
  1. Brain--Tumors. 2. Tumor markers. I. Staab, Adrian. II. Hagemann, Carsten. III. Title.
  [DNLM: 1. Brain Neoplasms--genetics. 2. Brain Neoplasms--metabolism. 3. Cell Hypoxia. 4. Gene Expression Regulation, Neoplastic. 5. Tumor Markers, Biological--genetics. 6. Tumor Markers, Biological--metabolism. WL 358 S132b 2010]
  RC280.B7S25 2010
  616.99'481--dc22
                                                        2010016663

Published by Nova Science Publishers, Inc. ✤ *New York*

Contents

Preface		vii
Introduction and Background		ix
Chapter 1	Biochemistry of Human Brain Tumor Cells	1
Chapter 2	Glycolytic Regulation in Human Brain Tumors	9
Chapter 3	Hypoxia Induced HiF-1 Gene Regulation in Human Glioblastoma	11
Chapter 4	Regulation via other Hypoxia Gene Regulators	25
Conclusions		27
Acknowledgments		29
References		31
Index		47

Preface

Hypoxia is an important phenomenon possessing a significant correlation with tumour progression, treatment result(s) and the overall disease prognosis.

Tumor oxygenation state leads to a series of genomic changes, enabling tumour cells survival or overcoming the oxygen deficient environment conditions.Tumour tissue growth requires a sufficient oxygen and nutrients supply. Tumor cell responses to hypoxic stress include adaptive proteomic changes allowing the cells to overcome nutritive deprivation or to escape their hostile environment by proliferation, invasion, or metastatic spread. Tumor cell proliferation requires rapid synthesis of macromolecules including lipids, proteins, and nucleotides.

Hypoxia-inducible factor 1 (HIF-1) is a multi-subunit proteinthat regulates transcription at hypoxia response elements (HREs)under hypoxic oxygenation conditions, while under normoxic oxygenation conditions, HIF-1α protein is subject to rapid degradation bya pVHL-mediated ubiquitin-proteasome pathway,while underit hypoxia blocks degradation leading to accumulation and translocation to the cellular nucleus and its binding together with HIF-1β on the so called Hypoxia responsive element within the hypoxia induced genes promoter region and thereby regulating the expression of the hypoxia responsive genes which proteins are expressed in human brain cancer especially in low grade astrocytoma and glioblatoma in a tumor stage and oxygenation specific manner. Genes which proteins are expressed in this manner are considered tumor marker genes as well as tumor therapeutic targets. Among the most important are Carbonic anhyrase 9 (CA9), N myc- Downregulated gene 1(NDRG1), Osteopontin (OPN), Vascular endothelial Growth Factor (VEGF) and Erythropoitin (EPO). HIF-1α is an important or the only regulator of these proteins under hypoxic conditions. CA9 is one of the most strongly hypoxia-inducible proteins but its expression pattern is not only related to its transcriptional induction by hypoxia in brain tumors, but also to effects of adverse microenvironmental stresses (such as diminished levels of glucose and bicarbonate. NDRG1 was shown at the same levels *in vivo* in normal human brain and human low-grade astrocytoma (WHO grade 2), while it showed a higher NDRG1 overexpression level in glioblastoma than in lowgrade astrocytoma.OPN was displayed a high level of cancer tissue expression specificity due to the favour expression in human GBM when compared to (LGA) based on the relative mRNA expression of hypoxia-related genes data. Regulation of OPN, mRNA and protein expression as a response to the hypoxic development in the tumor cell enviroment *in vitro* and *in vivo* represent an absolute phenomenon in human glioblastoma as a cell-specific post-transcriptionally regulated event. VEGF is a powerful hypoxia induced mitogen for endothelial cell growth and plays a critical role in the development of tumor vessels.OPN is a tumor-associated phosphoglycoprotein, has been described to be prognostic for tumor progression and survival in a number of solid neoplasms and linked to a "metastatic phenotype (Epo) is overexpressed at the mRNA level in human and mouse brain. Modulation of glycolysis reduces the hypoxic accumulation of HIF-1α protein in human tumor cells through a translational or post-translational process. Therefore, manipulation of tumor glucose levels represents a potential approach to therapeutically target HIF-1α.

In malignant glioma therapy, the main aim, as with all cancers, is to either eradicate the tumor or convert it into a controlled, quiescent chronic disease. Angiogenesis and hypoxia induced, HIF-1α regulated genes inhibition remains the main parts of therapeutic approaches in human oncology. It is well known that cancer cell metabolism can be perturbed specifically at the level of glycolysis leading to interesting therapeutic activities in cancer that can be displayed.

Introduction and Background

Hypoxia is an important phenomenon possessing a significant correlation with tumour progression, treatment result(s) and the overall disease prognosis [Wang GL 1995 Kallio PJ 1997], representing an important tumour microenvironment factor that significantly influences tumour cells' behaviour via activation of genes encoding proteins involved in hypoxic stress adaptation. Tumor hypoxia is an important cancer prognosis indicator associated with aggressive growth, metastasis, and poortreatment response [Höckel M 1996, Brizel D M 1996],with further association with malignant progression and human cancer's poor outcome. The tumor'soxygenation state leadsto a series of genomic changes, enabling tumour cell survivalfor overcoming the oxygen deficient environment conditions. Hypoxia tumour tissue growth requires a sufficient oxygen and nutrients supply. Tumor cell responses to hypoxic stress include adaptive proteomic changes allowing the cells to overcome nutritive deprivation or to escape from their hostile environment by proliferation, invasion, or metastatic spread [Fandrey J, 1995]. However, proliferating tumour cells quickly overgrow the oxygen diffusion distance from the nearest blood vessel (100 - 150 µm), leading to a highly irregular tumour vasculature, with arteriovenous shunts, blind ends, and incomplete endothelial linings. Blood flow, as a consequence, becomes less efficient than in normal tissues [Richard DE 1999]. Hypoxia-inducible factor 1 (HIF-1) is a multi-subunit proteinthat regulates transcription at hypoxia response elements (HREs)and is composed of 2 basic helix-loop-helix proteins: the α subunit,HIF-1α, and the constitutively expressed HIF-1β (alsoknown as aryl hydrocarbon receptor nuclear translocator [ARNT]), during normoxia HIF-1α is hydroxylatedon several proline and asparaginyl residues, which enables high-affinitybinding of HIF-1α to von Hippel – Lindau tumor suppressorprotein (vHL), a component of a ubiquitin ligase complex thatubiquitinates and thereby targets HIF-1α for proteosomal degradation.Under normoxic oxygenation conditions, HIF-1α protein is subject to rapid degradation process by pVHL-mediated ubiquitin-proteasome pathway, Under hypoxic conditions the O_2-dependent hydroxylation of HIF-1αis decreased, which prevents its degradation,whereas hypoxia blocks its degradation leading to accumulation and translocation to the cellular nucleus and its binding together with HIF-1β on the so called Hypoxia responsive element within the hypoxia induced genes promoter region [Huang LE 1996, Kallio PJ 1997].

The association of HIF-1α with pVHL is triggered by the post-translational hydroxylation of proline residue that is mediated by prolyl hydroxylase (PHD) or HIF prolyl hydroxylase (HPH). The hypoxia-inducible factor (HIF-1) is an oxygen-dependent transcriptional activator, which plays crucial roles in the angiogenesis of tumours and mammalian development. HIF-1 consists of a constitutively expressed HIF-1β subunit and one of three subunits (HIF-1α,

HIF-2 α or HIF-3α). The stability and activity of HIF-1α are regulated by various post-translational modifications:hydroxylation, acetylation, and phosphorylation. Therefore, HIF-1α interacts with several protein factors including PHD, pVHL, ARD -1, and p300/CBP. Under normoxia, the HIF-1α subunit is rapidly degraded via the von Hippel - Lindau tumor suppressor gene product (pVHL)- mediated ubiquitin-proteasome pathway. The association of pVHL and HIF-1α under normoxic conditions is triggered by the hydroxylation of prolines and the acetylation of lysine within a polypeptide segment known as the oxygen-dependent degradation (ODD) domain. On the contrary, under hypoxic conditions, the HIF-1α subunit becomes stable and interacts with its co-activators such as p300/CBP to modulate its transcriptional activity. Eventually, HIF-1 acts as a master regulator of numerous hypoxia-inducible genes under hypoxic conditions. The target genes of HIF-1 are especially related to angiogenesis, cell proliferation/survival, and glucose/iron metabolism. Moreover, it was reported that the activation of HIF-1α is closely associated with a variety of tumors and oncogenic pathways. The blocking of HIF-1α itself or HIF-1α interacting proteins inhibit tumor growth. Based on these findings, HIF-1 can be a prime target for anticancer therapies. It has been shown that carbonic anhydrase IX (CA IX) represents an important intrinsic marker of hypoxia. Carbonic anhydrase IX (CA IX) is one of the most strongly hypoxia-inducible proteins. Due to its different characteristics it is considered an intrinsic marker of hypoxia. Carbonic anhydrase IX (CA - IX) belongs to the carbonic anhydrase family ofmetalloenzymes that catalyse the reversible hydration of carbon dioxide to carbonic acid and play important roles in various biological processes related to acid-base balance [Chomczynski P 1987].

Several genes are regulated via the hypoxia induced HIF-1α activation. Among these important genes, is CA IX, which is a transmembrane N-glycosylated isoenzyme localised at the cell surface ina trimer formcomposed of monomeric subunits of 58/54 kDa [Pastorek J 1994]. The large CA IX extracellular molecular part contains an N - terminally located proteoglycan-like region that is missing from other CAs and it possessesadhesion capacity [Závada J 2001, Wingo T 2000]. The central carbonic anhydrase domain exhibits high enzymatic activity and has a structural predisposition to serve as a receptor site [Ivanov S 2001]. The intracellular CA IX C-terminus is linked to the extracellular part by a single hydrophobic transmembrane anchor [Pastorek J 1994, Opavsky R 1996]. CA IX is frequently present in different types of tumor cells and absent from their normal counterparts [Závada J 1993]. Natural CA-IX expression occurs only in few normal tissues like the stomach epithelia, the small intestine and gallbladder [Pastoreková S 1997]. CA IX is a recognised target of HIF-1 transcriptional complex that is highly responsive to changes inthe oxygen levels in tumor cells *in vitro*and shows a typical hypoxic pattern of distribution in a wide variety of tumor tissues [Wykoff C 2000, Ivanov S 2001].

The CA IX expression pattern is not only related to its transcriptional induction by hypoxia, but also to effects of adverse microenvironmental stresses (such as diminished levels of glucose and bicarbonate) and to high protein stability in reoxygenated cells. This suggests that CA IX potentially also detects the tumor regions that have experienced hypoxia either alone or in combination with low glucose or low bicarbonate before the tissue removal. Therefore, CA – IX serves as a marker of actual intra-tumoral microenvironments. Also, CA - IX serves as a 'record' of the expired hypoxia and stresses related to hypoxia. CA-IX expression occurs in different tumour tissues where CA - IX expression isnormally absent like thebladder, kidney, breast, lung, head and neck,and cervix uteri, i.e. [Pastorekova S 2004] CA IXexpression also occurs in mouse brains [Hilvo M 2004]. For the human brain, CA IX expressionwas only shown in normal human braintissuewith slight or no expression in epithelial cells of the choroid plexus [Ivanov S 2001],and increasingly shown in high grade astrocytomas (Grade III and Grade IV) [Said HM 2007].

Afurther level of O_2-dependent regulation exists: the hydroxylationof an asparagine residue by factor inhibiting HIF-1α (FIH) blocksthe interaction of HIF-1α with p300/CBP transcriptional coactivatorproteins, thereby decreasing transcription of HIF-1α-regulatedgenes at normoxia. When HIF-1α levels increase in response tohypoxia in tissues, functional HIF-1 regulates transcriptionat HREs of target gene regulatory sequences, which results inthe transcription of genes such as Carbonic anhydrase 9 [Bunn HF1996, Fandrey J 1995].

Cellular pH level and carbon content influence expression of HIF-1α [Svastová E 2004], and CA-IX in tumour cells to overcome the stress situation by providing an energy source needed for their various activities [Svastová E 2004] and therefore contributeto the resistance of these cells towards radiation therapy and aretherefore resistant to radiation therapy since tumor hypoxia is well recognized as a major factor contributing to radioresistance.[Overgaard J 1989,Knisely J 2002, Koukourakis MI2006] The interference into this process through glycolysis inhibition[Nodin C 2005,Alabovskii VV2004]represents an important issue.

Vascular endothelial growth factor (VEGF) is a powerful hypoxia induced mitogen for endothelial cell growth and plays a critical role in the development of tumor vessels]. Osteopontin (OPN), a tumor-associated phosphoglycoprotein, has been described to be prognostic for tumor progression and survival in a number of solid neoplasms and linked to a ''metastatic phenotype Erythropoietin (Epo) has been reported to be overexpressed at the mRNA level in human and mouse brains"(Said HM 2007).

Hypoxia induced regulation ofanother hypoxia induced gene, NDRG1 was shown at the same levels *in vivo*in a normal human brain and human low-grade astrocytoma (WHO grade 2), while it showed a higher NDRG1 overexpression levels in glioblastoma than in lowgrade astrocytoma (Said HM 2009).

Glycolysis inhibiton is effective against cancer cells under hypoxic conditions, which is frequently associated with cellular resistance to conventional anticancer drugs and radiation therapeutic approaches.Otto Warburg demonstrated that ascites tumor cells had high rates of glucose consumption and lactate production despite sufficient oxygen availability necessary for complete glucose oxidization [Hedeskov CJ 1968]. The so called ''Warburg effect'' represents a metabolic hallmark of aggressive tumors; however, the phenotype is also observed in non-transformed cells during rapid proliferation [Hedeskov CJ 1968, Wang T, 1976].Glycolytic enzyme inhibitors functional actions are based primarily on ATP depletion including: A) Hypoxia-linked cancer-cell resistance reduction; B) Anabolic and Energetic processes inhibition; C);Improvement of drug selectivity by exploiting particular glycolysis dependency of cancer cells D) ATP-dependent multi-drug resistance reduction; E) cytotoxic synergism with conventional cancer treatments [Scatena R 2008]. Hypoxia is an adverse prognostic factor in many tumor entities and tumor hypoxia has been associated in particular with poor response to radiotherapy [Evans SM 2003, Grumann T 2006, Höckel M1993, Höckel 1996, Höckel M2001Mayer R2005, Molls M 1998, Nordsmark M2005, Rohrer Bley C2006, Vaupel P2006]. In order to improve the therapeutic efficacy of radiotherapy and to overcome the radioresistance of hypoxic tumors, a modification of tumor oxygenation and the targeting of hypoxia-related moleculeshasbeen investigated [Hagen T1975, Molls M 1998, Sakata K 2006]. The transcription factor hypoxia-inducible factor-1 (HIF-1) is one potential target molecule to improve the radio sensitivity of hypoxic tumors and consists of the two subunits HIF-1α and HIF-1β. It activates over 60 known genes regulating multiple functions relevant to cellular survival and proliferation in an oxygen-dependent manner [Lee JW 2004].HIF-1α degradation is O_2-dependent by the prolyl hydroxylases activity. HIF-1 is associated with a poor outcome for multiple cancer types and high HIF-1 expression is a predictor of poor prognosis after radiotherapy [Aebersold DM 2001, Birner P 2001, Bos R 2003, Giatromanolaki A 2001]. Cancer cell energy metabolism deviates from that of normal tissues by maintaining high glycolytic rates which has also been associated with disease progression in several tumor entities [Brizel DM 2001, Galarraga J, 1986 Isidoro A 2005]. Hypoxic accumulation of HIF-1α occurs in a cell-type-specific manner [Vordermark D 2004], and is strongly dependent on glucose availability [Sobhanifar S 2005, Vordermark D 2004].

As it could beseen in the direct inhibition ofa hypoxia induced gene like NDRG1 (Said HM 2009) via siRNA or indirect inhibition, interfering with the cancer cell glycolytic activities via application of IAA (Said HM 2007, Staab A 2007) might be a potential therapeutic tool for regulating the expression of this gene in glioblastoma. Detailed understanding of how hypoxia regulates transcription of the NDRG1 gene increases knowledge of the cellular responses of normal and cancer cells towards low oxygen tension.

Chapter 1

Biochemistry of Human Brain Tumor Cells

Metabolism in Human Brain Tumor Cells

Tumor cell proliferation requires rapid synthesis of macromolecules including lipids, proteins, and nucleotides. Many tumor cells exhibit rapid glucose consumption, with most of the glucose-derived carbon being secreted as lactate despite abundant oxygen availability (the Warburg effect). (DeBerardinis RJ, 2007). An ordered pattern of metabolic and morphologic changes occurs during neoplastic cell transformation. Apparently, an increase in the flux of certain metabolic pathways such as the hexose - monophosphate shunt and glycolysis develops during transformation of many cell types. This metabolic aberration is conventionally explained as a consequence of a higher metabolic requirement (Banash P 1986). Gliomas represent 50% of primary brain tumors, and their prognosis remains poor despite the advances in diagnosis and therapeutic strategies (Lamari F, 2008).Positron emission tomography is used to study abnormalities in the glucose oxidative metabolism in human cerebral gliomas. It was possible to see that tumor regional cerebral glucose consumption was not depressed and regional glucose extraction ratios were similar for both tumor and brain tissue, (Rhodes CG, 1983). Acidic pH1 decreased the Lactate / pyruvate ratio and ATP level while both were markedly increased by basic pH1 (Miccoli L, 1996). In some other experimental analysis, it was observed in human xenografted gliomas that the Lactate / pyruvate ratios increased 3 - 4 fold and HK activity was of 2 - 4 fold lower than that of normal rat brain tissue, used as the control. The mitochondria-bound HK (mHK) fraction varied considerably and represented 9 to 69% of the total HK of that normal rat brain (Oudard S, 1997). Furthermore, in another series of experiments, the Lactate to pyruvate ratio was >1, suggesting that the energy metabolism in LGG is glycolytic in nature, particularly in the tumor centre. Peripheral samples of tumors showed increased glucose consumption and cytochrome c - oxidase activity (Lamari F, 2008).

Hexokinase bind to a mitochondrial porin involving peripheral benzodiazepine receptors.

HK and peripheral benzodiazepine receptors inhibition by lonidamine and diazepam led to synergistic anti tumoral activity in xenografted gliomas. Co-inhibition of these two receptors will lead to a decrease in glycolysis, often elevated in these tumors, without modifying energetic metabolism of normal cells (Oudard S, Miccoli L, 1998).

Potential Role of Glycolytic Inhibitors in HumanBrain Cancer Treatment

Glycolysis, the transformation of glucose to pyruvate, is akey step for the acquisition of ATP in all mammalian cells,including cancer tissues. Glucose transporters are commonlyoverexpressed in human malignancies enhancing glucose influxin the proliferating cancer cells (Macheda ML 2005). Glycolytic inhibitors are particularly effective against cancer cells with mitochondrial defects or under hypoxic conditions, which are frequently associated with cellular resistance to conventional anticancer drugs and radiation therapy (Pelicano H, 2006). Energy metabolism is considered for brain cancer treatment through metabolic targeting in the normal orthotopic tissue. The glucose transporter, GLUT-1, is enriched in the brain capillary endothelial cells and mediates facilitated glucose diffusion through the blood brain barrier. Glucose is metabolized mostly, to pyruvate, which enters neurons and glia mitochondria and is converted to acetyl-CoA before entering the TCA cycle. In well-oxygenated normalcells, pyruvate enters the mitochondria where, by the enzymicactivity of pyruvate dehydrogenase, it is transformed to acetyl - Co A,the substrate for ATP production through the Krebs cycle (Harris RA 2002). Under normal conditions, 13% of glycolytic pyruvate is converted to lactate (Clarke DD 1999).Because intracellular acidosis triggers apoptosisblocking increased glycolytic activity by down regulating HIF-1α may reduce apoptosis of the hypoxic cells (Schmaltz C

1998). Within this context, results showed that iodide acetate minimized or inhibited HIF-1α protein and CA-IX protein and mRNA expression in glioblastoma cells under long term *in-vitro* hypoxia (Said HM 2007) as well as in other tumors of different origin (Staab A 2007).It has previously shown that CA IX may be a therapeutic target for cancer, since, inhibition ofcarbonic anhydrase isoenzymes with bacteriostatic or non bacteriostatic sulfonamides e.g. acetazolamideresult ineither reduced tumor invasiveness or blocked tumor growth, respectively (Svastová E 2004, Nodin C 2005). Furthermore, CA isoenzyme antagonism has been observed to augmentthe cytotoxic effects of various chemotherapeutic agents, includingplatinum-based drugs.

There is a clear involvement of glucose concentration in the HIF-1α and CA-IX expression regulation, since application of the glycolsis inhibitor iodide acetate lead to a minimized expression of both of them in different glioblastoma cell lines (Figure). Accumulation of HIF-1-α depends on Glucose concentration. Glycolysis inhibitors, when added under hypoxia, lead to a reduced accumulation of HIF-1α. It has also been shown that inhibition of mitochondrial respiration leads to the inhibition of HIF-1α stabilization at low O_2 concentrations [Mateo J 2003].Interference in the glycolysis path of glioblastomas by iodide acetate can represent a therapeutic alternative in CA-IX involved therapeutic approaches and potentially other HIF-1α regulated hypoxia induced genes. On the other hand, we can use CA-IX status for diagnostic purposes to potentially aid in the selection of patients who might benefit from CA-IX-targeted therapies (2003 Bui MH). Ca9 can represent an optimal target for therapeutic applications in hypoxia related glioblastoma tumours. In glioblastoma inhibition of HIF-1α regulated hypoxia induced genes like Ca9 is accomplished via the functional interference into the tumor cell glycolysis pathaway via IAA and Chetomin.Cancer cell energy metabolism deviates from that of normal tissues by maintaining high glycolytic rates which has also been associated with disease progression in several tumor entities [Brizel DM 2001, Galarraga J1986, Isidoro A 2005].We and others have recently shown that the hypoxic accumulation of HIF-1α occurs in a cell-type-specific manner [Said HM 2007, Staab 2007]and is strongly dependent on glucose availability [2005 Kwon SJ, Sobhanifar S 2005, Vordermark D 2005].Glucose metabolism modulation investigation using the glycolysis inhibitors iodoacetate (IAA) or 2-deoxyglucose (2 - DG), affects the hypoxic accumulation of HIF-1α and to the characterization ofthe mechanism of glucose-dependent HIF-1α regulation had been accomplished.The hypoxic accumulation of HIF-1α depends strongly on the availability of glucose [2005 Kwon SJ, Vordermark 2005].Recent work has shown that that there is strong evidence that HIF-1 regulates glucose metabolism and maintenance of tumor growth [Griffiths JR 2002, Yasuda S 2004].

One explanation forthis effect is increased oxygen availability for prolyl hydroxylation of HIF-1α when mitochondrial oxygen consumption is reduced such that hypoxia is not recognized by prolyl hydroxylases[Hagen T 2003].

Glycolysis inhibitors, when added under hypoxia, lead to a reduced accumulation of HIF-1α. The regulation of HIF-1α by inhibition of glycolysis is independent of the activation by prolyl hydroxylases in HT1080 cells. Furthermore, pyruvate was not able to increase hypoxic HIF-1α levels when glycolytic inhibitors were added to HT1080 cells, suggesting that the lack of pyruvate is not the reason for the reduced accumulation of HIF-1α underhypoxic conditions when glycolysis was inhibited. In contrast to the report by Lu et al. we found no evidence for a key role of pyruvate as a glycolytic metabolite promoting HIF-1α accumulation [Lu H 2002].

To further investigate the mechanism by which the availability of glucose or glucose metabolites interacts with HIF-1α, we performed real-time RT-PCR to quantify the mRNA expression of the HIF-1α gene in cells under normoxia or hypoxia treated with IAA or 2-DG. We did not observe any significant changes in the HIF-1α gene expression profiles after incubation of HT1080 cells with IAA or 2-DG. This finding corresponds to previous reports showing that hypoxia inhibits mRNA translation by suppressing multiple key regulators [Liu L 2004, Liu L2006] and limited nutrient availability can lead to drastically reduced protein synthesis [Faulhammer F 2005].

We therefore presume that glucose levels affect the expression of HIF-1α on a translational level or by phosphorylation instead of transcriptional regulation. A likely explanation for the reduced hypoxic accumulation of HIF-1α after inhibition of glycolysis, as compared to full glucose availability, is the glucose's dependence onmRNA translation.

The interaction of glycolysis and the HIF-1 pathway may well explain in part the effects of glycolysis inhibitors shown in preclinical and clinical studies where such agents have increased the efficacy in chemotherapy protocols and after radiation treatment [Maher JC 2004, Varshney R 2005]. One can speculate that this effect depends on a reduced intratumoral accumulation of HIF-1α and thereby reduced expression of HIF-1-regulated genes. Previous publications have shown that glucose deprivation leads to an activation of multiple signal transduction pathways, changes in gene expression and induction of oxidative stress which mediate glucose-deprivation-induced cytotoxicity and metabolic oxidative stress in human cancer cells[Yun H 2005, Ahmad IM 2005].

Modulation of glycolysis reduces the hypoxic accumulation of HIF-1α protein in human tumor cells through a translational or post-translational process. Therefore, manipulation of tumor glucose levels represents a potential approach to therapeutically target HIF-1α. A clear involvement of glucose availability in the hypoxic HIF-1a and CA IX expression in malignant glioma cells exist since application of the glycolysis inhibitor iodoacetate led to a sharply reduced expression of both proteins and CA9 mRNA in all cell lines tested. Most previous reports have focused on the effects of HIF-1 on glucose metabolism, rather than vice versa: glycolytic enzymes are induced by hypoxia and lactate productionbecauseglycolysis is a major cause of the acidic extracellular pH of tumors [Semenza GL 1994].

Suboptimal oxygen availability switches on cellular metabolism to anaerobic pathways for ATP production, which occurs through pyruvate transformation to lactic acid via the catalytic activity of lactate dehydrogenase 5 [Holbrook JJ 2002, Harris RA 1975].Because intracellular acidosis triggers apoptosis blocking, increased glycolytic activity by down regulating HIF-1a may reduce apoptosis of hypoxic tumor cells [Schmaltz C1998]. We could previously show in non-brain tumor cell lines that the hypoxic accumulation of HIF-1a and expression of CA IX in vitro depend on the glucose concentration in the medium [Vordermark D 2005, Vordermark D 2005].

Glycolysis inhibitors, when added under hypoxia, led to a reduced accumulation of HIF-1α via a translational or post-translational effect [Staab A 2007]. Interference with the glycolysis of malignant gliomas by iodide acetate may therefore represent a therapeutic approach in targeting HIF-1a or CA IX. HIF-1 inhibition has been shown to slow tumor growth in in-vitro and in-vivo tumor models [Kung AL 2004, Welsh S 2004] and to act synergistically with other treatment modalities such as radiotherapy [Moeller BJ 2004].

In human fibrosarcoma cell line that HIF-1 targeting with chetomin (150 nM) suppresses the transcriptional response to hypoxia and reduces hypoxic radioresistance *in vitro*. This is, to our knowledge, the first report of increased radiosensitivity of hypoxic cells in vitro in response to chetomin. In a human fibrosarcoma cell linewithHIF-1 targeting, chetomin (150 nM) suppresses the transcriptional response to hypoxia and reduces hypoxic radioresistance in vitro (Staab A 2007). Experimental conditions now chosen were based on our previous observation that a near-maximal HIF-1α expression occurs at 12 h of hypoxia at an oxygen concentration of 0.1% O_2, which represents a level of hypoxia that is frequently observed in solid tumors and is radiobiologically relevant [Vordermark D 2004]. At the dose level of 150 nM, chetomin exhibited a maximum specific effect on HRE-regulated elements.

Potential Role of Chemical Inhibitors in Human Brain Cancer Treatment

Most anticancer drugs are transported by either active transport or passive diffusion into cells, where they frequently undergo further metabolism (Stubbs M 2000).

It is well known that Carbonic anhydrase isoform IX (CA IX) is highly over expressed in many types of cancer. Its expression, which is regulated by the HIF-1α(Wykoff CC2000, Said HM 2007) transcription factor, is induced by hypoxia and correlates with a poor response to classical anti cancer therapeutic approaches like chemo- and radiotherapies (Said HM 2007). This chemo- and radio resistance occurs due to the CA IX contribution to the tumor environment acidification by efficiently catalyzing the hydration of carbon dioxide to bicarbonate and protons leading to metastatic phenotypes acquisition and chemoresistance to weakly basic anticancer drugs (Supuran CT 2000). Among them are CA-IX-selective inhibitors, whichcan be inhibited via potent inhibitors derived from acetazolamide, benzenesulfonamides and ethoxzolamide which have been shown to inhibit the growth of several tumor cells*in vitro* and *in vivo*(Supuran CT 2006), (Supuran CT 2007).These drugs are pH sensitive.Therefore it is suggested that their cytotoxic activity depend on both intracellular pH (pHi) and pHe (CT-2008). Targeting CA IX with suchspecific CA IX inhibitors (Pastorekova S 2004), or also antibodies (Chrastina A 2003)should contribute, on one hand, to the enhancing action of weakly basic drugs and on the other hand, to reduce the acquisition of metastatic phenotypes by controlling the pH imbalance in the tumor cells. (Thiry A, 2008).Selective CA IX inhibitors could prove useful for elucidating the role of CA IX in hypoxic cancers, for controlling the pH imbalance in tumor cells and for developing diagnostic or therapeutic applications for tumor management. Ca9 specific enzymatic inhibition activity by specific inhibitors belonging to the group of sulphonamides like indisulam, reverts these processes, establishing a clear-cut role for CA IX in tumorigenesis. (Thiry A, 2008).Practically, CA inhibitors have been previously shown to elicit synergistic effects when used in combination with other chemotherapeutics agents in animal models (Teicher BA 1993). The antiproliferative effect of CA inhibitors might also be due to their effect on other CA isoforms such as CA II or CA V, which provide the bicarbonate substrate for cell growth in carboxylation reactions involved in lipogenesis, nucleotide biosynthesis and gluconeogenesis, among others, thereby limiting the unrestrained proliferation of the tumor cells (Scozzafava, A 2006), (Supuran CT 2007) and (Supuran CT 2003).

Inhibition of this enzymatic activity by specific inhibitors, such as the sulfonamide indisulam reverses these processes, establishing a clear-cut role for CA IX in tumorigenesis. Thus, selective CA IX inhibitors could prove useful for elucidating the role of CA IX in hypoxic cancers, for controlling the pH imbalance in tumor cells and for developing diagnostic or therapeutic applications for tumor management. Indeed, fluorescent inhibitors and membrane-impermeable sulfonamides have recently been used as proof-of-concept tools, demonstrating that CA IX is an interesting target for anticancer drug development. The research of these CAIX inhibitors is undergoing continuous development. One of the mean reasons for that is in order to reach a high selectivity of these drugs and to avoid any side effects by other CA isozymes inhibitionthusavoiding having them play their physiological roles (Thiry A, 2006).

Glucose Metabolism in Human Cancer Cells

Cellular glucose metabolism may occur either aerobically or anaerobically. In aerobic metabolism, glucose is converted to CO_2 and H_2O via the tricarboxylic acid (TCA) cycle with the generation of about 36 moles of ATP per mole of glucose consumed. In anaerobic glycolysis, glucose is metabolized to lactic acid, producing 2 moles each of ATP and H+ ions per mole of glucose [Stryer L, 1988]. For fundamental thermodynamic reasons [Pfeifer T 2001], the efficiency of aerobic metabolism is achieved at the cost of decreased maximum rate, and ATP production by the respiratory pathway rapidly saturates at high levels of glucose or limited oxygen supply. In the lower-yield anaerobic pathway, more of the energy from glucose degradation is used to drive the reaction, allowing a greater maximum rate of metabolism. The net ATP production rate of the anaerobic pathway, in the presence of adequate glucose, can be similar to that of the aerobic route despite the relative inefficiency.

Malignant brain tumors from either humans or animal models lack metabolic flexibility, in contrast to a normal brain that oxidizes glucose as well as ketone bodies for energy. They are largely dependent on glucose for energy [Seyfried TN 2003, Mangiardi JR 1990].Rhodes CG 1983, Nagamatsu S 1996, Roslin M 2003,Floridi A 1989, Galarraga J 1986, Mies G 1990, . Oudard S 1997]. Enhanced glycolysis produces excess lactic acid that can return to the tumor as glucose through the Cori cycle [Tisdale MJ 1997].

Normal mammalian cells under physiological conditions utilize high-yield aerobic glucose metabolism, but can adapt to periods of hypoxia by elevating the anaerobic pathway, provided the transition to hypoxia is gradual and allows for induction of response mechanisms such as HIF. The energy cost of this transition is substantial, as the output of ATP per mole of glucose is reduced by over 90%. To compensate this decreased efficiency, glycolytic flux must increase several-fold. Warburg [WARBURG O 1930] first demonstrated tumour glucose metabolism alteration. Transformed cells *in vivo* and *in vitro* typically rely on anaerobic pathways to generate ATP from glucose even in the presence of abundant oxygen. A rough correlation between malignancy degree and glycolytic rate has long been noted [Burk D 1967]. The decreased efficiency of anaerobic metabolism is compensated by increased glucose, flux, maintaining energy production sufficiently in excess of basal metabolic demands to allow for cellular proliferation.

Metabolic Control Analysis in Human Brain Tumors

Metabolic control analysis evaluates the degree of flux in metabolic pathways and can be used to analyze and treat complex diseases [Veech RL 2004, Greene AE 2003]. The approach is based on findings that compensatory genetic and biochemical pathways regulate the tumor cells' phenotype and bioenergetic potential [Veech RL 2004, Greene AE 2003 and Strohman R 2002]. As rate-controlling enzymatic steps in biochemical pathways are dependent on the physiological system metabolic environment, the management of disease phenotype depends more on the flux of the entire system than on the expression of any specific gene or enzyme alone [Strohman R 2002, Kacser H 1981, and Greenspan RJ 2001]. Complex disease phenotypes can be managed through self-organizing networks that display system wide dynamics involving glycolysis and respiration. Global manipulations of these metabolic networks can restore orderly adaptive behavior to widely disordered states involving complex gene-environmental interactions [Strohman R 2002, Kacser H 1981, Greenspan RJ 2001, Seyfried TN 2003, Seyfried TN 2005].

Chapter 2

Glycolytic Regulation in Human Brain Tumors

Involvement of Glucose Availability in Hypoxia Induced Gene Expression

Glycolytic enzymes are inducedby hypoxia and lactate production by glycolysis is a major cause of the acidic extracellular pH of tumors[Semenza GL 1994]. Suboptimal oxygen availability switches on cellular metabolism to anaerobic pathways for ATP production, which occurs through pyruvate transformation to lactic acid via the catalytic activity of lactate dehydrogenase 5 [Harris RA 1975, Holbrook JJ 2002]. Because intracellular acidosis triggers apoptosis blocking, increased glycolytic activity by down regulating HIF-1a may reduce apoptosis of hypoxic tumor cells [Schmaltz C 1998]. We could previously show in non-brain tumor cell lines that the hypoxic accumulation of HIF-1a and expression of CA IX in vitro depend on the glucose concentration in the medium [Vordermark 2005,Vordermark D, Kraft P 2005]. Glycolysis inhibitors, when added under hypoxia, led to a reduced accumulation of HIF-1α via a translational or post-translational effect [Staab A 2007]. Interference with the glycolysis of malignant gliomas by iodide acetate may therefore represent a therapeutic approach alternative. In glioblastoma there is a clear involvement of glucose availability in hypoxic HIF-1α and CA IX expression in malignant glioma cells since application of the glycolysis inhibitor iodoacetate led to a sharply reduced expression of both proteins and CA9 mRNA in all cell lines tested. Most previous reports have focusedon the effects of HIF-1 on glucose metabolism, rather than vice versa.

Chapter 3

Hypoxia Induced HiF-1 Gene Regulation in Human Glioblastoma

Hypoxia Induced Ca9 Expression in Human Brain Tumors Cells

Endogenous hypoxia markers are genes or gene products that are specifically up-regulated under hypoxic conditions. Tumor hypoxia has been recognized to confer resistance to anticancer therapy since the evolvement of this branch of cancer therapy and its fundamental role in tumorigenesis has been established. Hypoxia-inducible factor HIF–1α has been identified as an important transcription factor that mediates the cellular response to hypoxia, promoting both cellular survival and apoptosis under different conditions, and has been extensively studied as an endogenous hypoxia marker and its mechanism of accumulation under hypoxia is well understood (Harris AL 2002, Semenza GL 2003).

Ca9 that is regulated by HIF-1α (Figure 1) has been shown to be induced in a wide range of malignant cells, *in – vitro*(Wykoff CC2000),in a high degree of expression that overlaps with pimonidazole (Olive PL 2001, Beasley NJ2001).

Also, Ca9 has been shown to be over expressed in many tumorswhich isa common feature of the cancer cells required for tumor progression. It may contribute to a tumor microenviroment bymaintaining extracellular acidic pH and helping cancer cells to grow (Ivanov S 2001).

Most studies of tumor hypoxia have been based on direct pO_2 measurements using oxygen microelectrodes (Evans SM 2000). However, there are limitations due to the invasive nature of the procedure. Hypoxia marker drugs, such as nitromidazole derivatives (e.g., EF5 and pimonidazole), also have been used to detect hypoxia in tumor samples; however, the requirement of prior administration restricts their clinical use to prospective studies (Kaanders JH 2002, Brown JM 2002).

Pimonidazole hydrochloride was selected as thehypoxia marker because of its high water solubility, chemicalstability, efficient tumor uptake, and low toxicity. 2000 mg/m^2 of pimonidazole hydrochloride, 1-[(2-hydroxy-3-piperidinyl) propyl]-2-nitroimidazolehydrochloride (Hypoxyprobe) in 100 ml normal saline i.v.over 20 min is the maximum tolerated dose (Kennedy AS 1997) on patients.

The *in-vitro* data presented suggests that hypoxia may contribute to increased levels of CA-IX expression in gliomas, *in - vitro*. Western and northern blot analysis revealed a difference between four cell lines in an expression level under the various O_2 conditions but with almost a similar tendency in its expression behaviour. One of the factors of these differences together with the different response to reoxygenation could be explained by the genetic background of the cell lines investigated (Said HM, 2007)as it might indirectly influence at least their expression degree.

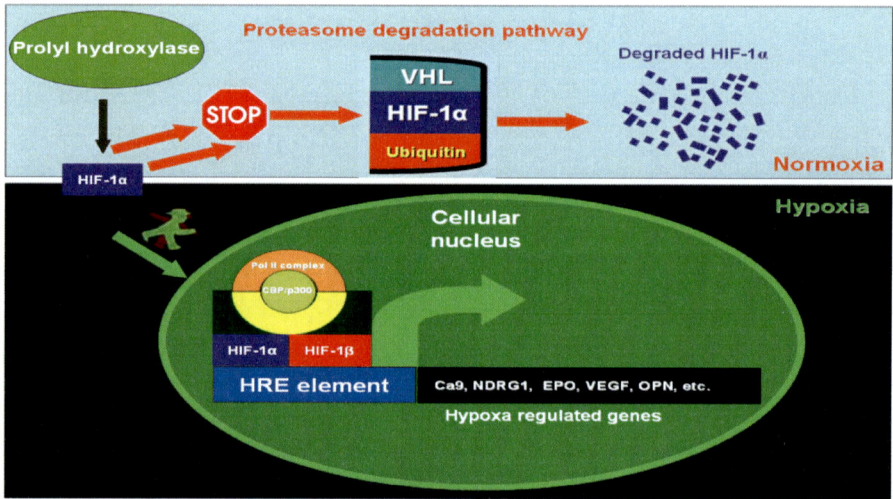

Figure 1. HIF-1α Induced regulation of hypoxia induced genes figure.
Under normoxia, HIF-1α is rapidly degraded via the *von Hippel – Lindau tumour suppressor gene product* (pVHL) – mediated ubiquitin proteasome pathway. When the tumor environment develop tohypoxic aerationconditionc, HIF-1α subunit becomes stable and interacts with coactivators of which its transcription machinery is consisted such as p300 / CBP to modulate the transcriptional activity of numerous hypoxia inducible genes, like carbonic anhydrase 9 (Ca9), OPN, NDRG1, EPO and VEGF and about 60 other hypoxia induced genes.

Also, a difference appeared in the expression tendency between CA-IX protein and the corresponding Ca9 mRNA. One of the reasons might be that the different posttranscriptional processing causes some difference between CA9 mRNA and CA-IX protein levels under coressponding conditions or that the CA-IX protein level posttranscriptionally regulates a Ca9 mRNA (Bishop JM, 1987, Prendergast G C1989).

But, because CA IX is relatively stable because of the strong basal expression with differences in inter-glialCA-IX expression and a high stabilityduring hypoxia / reoxygenation cycles(Swinson DE 2003,Giatromanolaki A 2001), therefore, when applicable*in-vivo*,we can say that CA IX expression status may be a reliable marker for glial tumor aggressiveness prediction associated with tumor hypoxia, so that it can be detected viaroutine clinical biopsies without the need for invasive procedures or prior drug administration.

Tumor cells and their survival and propagation can be further enhanced by genomic changes such as loss of apoptotic potential. These new cell variants have advantages over less adapted cells in a hypoxic microenvironment and expand through clonal selection becoming the dominant cell type in most cases. These variants further intensify hypoxia, establishing themselves through clonal selection, often becoming the dominant cell type. Further intensification of hypoxia, establishing a vicious circle of hypoxia, malignant progression, and treatment resistance is the normal consequence (Kim SJ 2004);glycolytic enzymes are induced by hypoxia (Semenza GL1994). The demonstration that an extracellular CA is up-regulated by microenvironmental tumor hypoxia has potentially important implications for understanding the regulation of tumor pH and the response to hypoxia. It has been widely held that lactate production by glycolysis is a major cause of the acidic extracellular pH of tumors.The hypoxic tumor environment favours the intensification ofanaerobic metabolic pathways in cancer cells. Suboptimal oxygen availability switches on cellular metabolismto anaerobic pathways for ATP production, which occurs throughpyruvate transformation to lactic acid via the catalytic activityof lactate dehydrogenase 5 (Holbrook JJ 1975).

Role of VEGF in Human Brain Tumors under Hypoxic Conditions

VEGF was originally discovered as a VPF, a protein secreted by tumour cells that potently stimulates ascites formation and vascular leak [Senger DR 1983]. Elevated levels of circulating VEGF provide a predictive measure of the progression of cancer and metastasis in certain cancers [Kuroi, K 2001]. In malignant gliomas, rapid cellular proliferation results in hypoxic conditions within the tumor. The release of humoral factors that promote angiogenesis, such as vascular endothelial growth factor (VEGF), seem to play a particularly important role in the process of neovascularization in malignant gliomas [Plate KH 1992, Kurimoto M1996].

The VEGF family of growth factors is unique, as it comprises the only angiogenic factor that also potently induces vascular leak. Although VEGF-induced angiogenesis is often accompanied by a vascular permeability response, VEGF-induced vascular leak is not required for angiogenesis [Eliceiri B 1999].

Angiogenesis is associated with expression of hypoxia-inducible factor HIF-1α and vascular endothelial growth factor (VEGF) in perinecrotic pseudopalisading glioma cells [Fisher, Zagzag 2005]. A powerful hypoxia-induced mitogen for endothelial cell growth is vascular endothelial growth factor (VEGF), which plays a critical role in the development of tumor vessels [Yancopoulos GD2000]. VEGF expression is stimulated by hypoxia regulation and oncogenic mutations. Inhibiting tumor angiogenesis provides an opportunity for therapy [Schlaeppi JM1999]. One of the key molecules responsible for the regulation of angiogenesis is vascular endothelial growth factor (VEGF), which is an endothelial cell-specific mitogen and a survival factor. VEGF also stimulates vascular permeability and recruits progenitor endothelial cells from bone marrow. Clinical observations have demonstrated that VEGF status is significantly correlated with neovascularization grade and prognosis in various types of solid tumors. It has also been shown that VEGF status is predictive of the resistance to various treatments, including radiotherapy, chemotherapy and hormonal therapy. Similarly, VEGF expression level in the tumor tissue was a significant predictor of relapse-free survival and overall survival in node-negative breast cancer patients treated with loco-regional radiotherapy [Linderholm B1999]. An important limiting factor in the growth rate of a tumor is its blood supply and that by interrupting new vessel formation, tumor growth can be effectively arrested. With the identification of various pro- and antiangiogenic factors and their signalling mechanisms, a better understanding of the molecular basis of angiogenesis has emerged [Davis GE 1995, Ingber DE 1995, Jang YC 1999, Plate KH 1992]. Cancer mortality has changed little over the past forty years, mainly because of our failure to develop curative chemotherapy for the common solid cancers. The way forward is to carry out extensive phase I and II clinical trials of the many new types of anticancer agents that have become available as a result of increased knowledge about cancer cells and how they differ from normal tissues. Tumor growth could be abrogated by inhibiting angiogenesis representsa major departure from the then prevalent concept of targeting tumor cells directly as a means of preventing their growth [1971Folkman J]. Different approaches to study VEGF by several study groups has been performed. Importantly, hypoxia has been thought of as a primary trigger that tips the balance toward angiogenesis, as it will increase HIF1 activity and hence VEGF levels [Laderoute KR 2000, MazureNM 1996, Diaz-Gonzalez JA2005]. In tumour lines expressed VEGF at low levels under normoxia, but several hours of hypoxia significantly increased VEGF expression *in vitro*. In contrast to Holash's theory, [Haroon ZA and his coworkers] did not observe vascular stasis before angiogenesis onset, but angiogenesis was accelerated when HIF1α was upregulated. They defined a physiological m called the 'acceleration model', where they proposed that HIF1α upregulation accelerates tumour angiogenesis, as opposed to its having a role in the initiation stages [Haroon ZA 2000]. VEGF upregulation is a trigger for initiation of angiogenesis from dormant metastases. Hypoxia, on the other hand, will increase HIF1 activity to upregulate VEGF while downregulating thrombospondin, to create a pro-angiogenic environment [Laderoute KR 2000]. Some metastatic tumours replace normal parenchymal cells with tumour cells thereby co-opting normal vasculature. For example, human breast tumours that have metastasized to liver show this phenotype. They have been reported to exhibit little evidence for hypoxia (indicated by expression of the endogenous hypoxia marker protein carbonic anhydrase IX) and express low levels of VEGF [Stessels, F2004, Colpaert CG2003]. Hepatic metastases of colorectal cancer, by contrast, show high levels of carbonic anhydrase IX and increased VEGF and clearly exhibit angiogenesis [Stessels, F2004]. VEGFR-1 antagonists and a tyrosine kinase inhibitor of VEGFR-2 reverted radiationrefractory tumor models to a radiation-sensitive phenotype. These findings suggest that the high VEGF expression might define a radio-resistant phenotype [Geng L 2001].

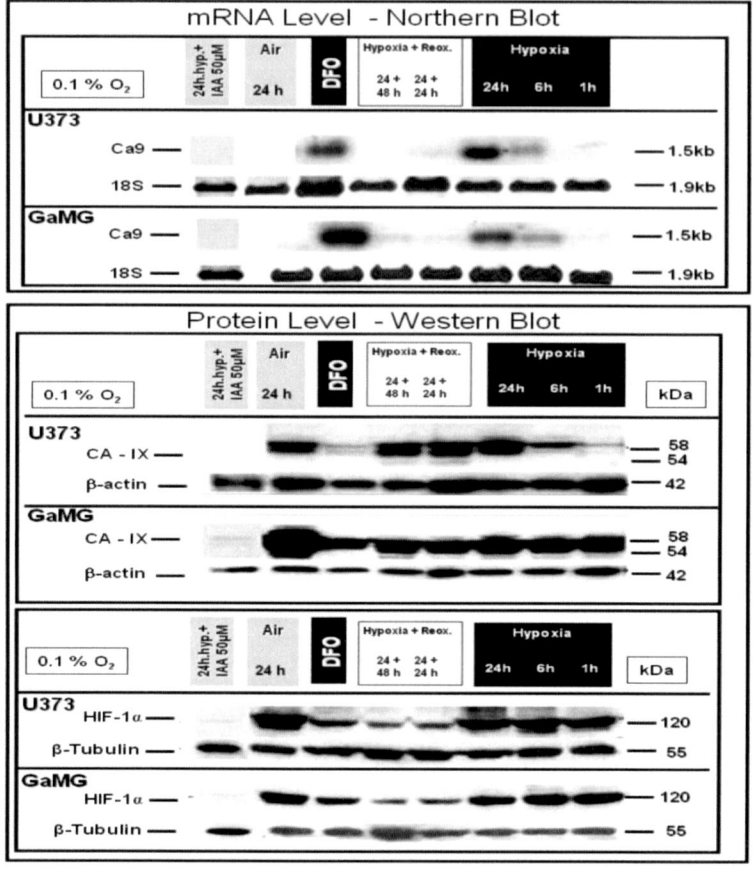

Figure 2. Ca9 and HIF-1α gene expression response to hypoxia in human brain tumor Cells *in vitro*.
Upper panels: Comparative Northern blot analysis of CA9 mRNA expression under in vitro hypoxia and reoxygenation in the human malignant glioma cell lines U373 and GaMG. Treatment with 100 μM DFO under aerobic conditions served as a positive control, 50 μM IAA was used as a glycolysis inhibitor. Representative Northern blots of CA9 mRNA with 18S as a loading control.
Lower panels: A-Comparative Western blot analysis of CA IX protein expression under *in vitro* hypoxia and reoxygenation in whole-cell lysates of the human malignant glioma cell lines U373 and GaMG. Treatment with 100 μM DFO under aerobic conditions served as a positive control, 50 μM IAA was used as a glycolysis inhibitor. Representative Western blots of CA IX protein with β-actin as a loading control. B-Comparative Western blot analysis of HIF-1a protein expression under *in vitro* hypoxia and reoxygenation in nuclear extracts of the human malignant glioma cell lines U373 and GaMG. Treatment with 100 μM DFO under aerobic conditions served as a positive control, 50 μM IAA was used as a glycolysis inhibitor. Representative Western blots of HIF-1a protein with β-tubulin as a loading control.

Hypoxia Induced OPN Expression in Human Brain Tumors Cells

Hypoxia regulated expression of OPN has been shown on several occasions [Said HM 2005, Said HM 2006, Said HM, Staab A 2007,Vordermark D 2006, Bache M 2006].Several previous reports have identified links between cancer outcomes and the level of HIF-1a protein or the expression of one or two individual genes that are induced by hypoxia, such as Carbonic Anhydrase IX[Said HM 2007, Vordermark D 2004,Vordermark D 2006]. OPN has displayed a high level of cancer tissue expression specificity due to the favour expression in human (GBM) when compared to (LGA) based on the relative mRNA expression of hypoxia-related genes data. Regulation of OPN, mRNA and protein expression as a response to the hypoxic development in the tumor cell enviroment *in vitro* and *in vivo* represent an absolute phenomenonin human glioblastoma as a cell-specific post-transcriptionally regulated event. Further, the phenotype of combined OPN and CA9 overexpression represent a clear phenotype that is associated with GBM, that also appears at a lower frequency in LGA. CA9 expression in GBM occurs at a high frequency both on the protein and mRNA level, as is the case for OPN, rendering OPN and CA9 as optimal diagnostic markers or targets for tumor-specific treatment approaches. Therapeutic strategies for treatment of human astrocytic tumors involving OPN, CA9 and to a certain extent EPO as a target molecule represent potential approaches in conjunction with tumor hypoxia in the human brain. Also, we have to finallymention that understanding of the prognostic value of the gene expression patterns we have identified and developing a small panel of well characterized markers that can be rapidly analyzed in clinical laboratories will be important forunderstanding what specific hypoxia-driven biological processes underlie the phenotypic differences between tumors in the high-hypoxia response group and those in the low-hypoxia response group.

Hypoxia Induced Epo Expression in Human Brain Tumors Cells

Erythropoietin (Epo) has been well characterized as a renal glycoprotein hormone that promotes erythrocytic progenitors survival, proliferation, and differentiation of hemopoietic tissues. Recombinant human Epo (rHuEpo) and related compounds have proved the most useful for treatment of the anemia associated with chronic renal failure and, more restrictedly, certain types of nonrenal anemias [Molineux G 2003, Heidenreich S 1991]. It has been shown in the research work of one group that 81% of human lung carcinoma tissues possessed Epo-binding sites as detected by use of biotinylated rHuEpo. Epo-R transcripts and Epo-R protein were subsequently demonstrated in human renal carcinoma, [Westenfelder C 2000]tumorsof the cervix and other organs of the female reproductive tract,[2003 Acs G, Yasuda Y 2001, Yasuda Y 2002] and in various specimens of common pediatric tumors such as neuroblastomas, brain tumors, hepatoblastomas, and Wilms' tumors[Batra 2003].By immunohistochemistry, Epo-R has been shown to be expressed in breast carcinoma[Acs G 2002, Arcasoy MO 2002, Hengartner MO 1994] and in vestibular schwannoma [Dillard DG 2001].

The role of Epo in tumor therapy needs to be further explored. Anemia-associated tissue hypoxia promotes angiogenesis, growth, and metastasis of tumors [Shannon AM 2003, Vaupel P 2004].

Since rHuEpo was introduced as a drug for treatment of renal anemia almost 20 years ago, several groups of investigators have carefully studied whether Epo can induce or promote tumor growth. Clinically, no evidence has been reported so far indicating that the erythrocytic growth factor Epo directly stimulates tumor cell proliferation. In addition, an elegant study in transgenic mice transfected with a construct that linked the human Epo gene to an erythroid-specific regulatory element has shown that the continuous stimulation of erythropoiesis leads to erythrocytosis but not to erythroleukemia [Madan A 2003] However, there is at least one case report of Epodependent leukemic transformation of myelodysplastic syndrome (MDS) to acute monoblastic leukemia (AML) [Bunworasate U 2001]. A careful examination has shown Epo-R expression on leukemia cells in 60% of patients with all French – American – British types of AML and in 29% of acute lymphoblastic leukemia (ALL) cases[Takeshita A 2002].

In vitro, a proliferative response to Epo was observed in 16% of patients. Patients with both Epo-R expression and in vitro response to Epo had shorter remission duration than those without Epo-R [Takeshita A 2002]. Thus, close observation for leukemic transformation is necessary in patients with MDS on rHuEpo therapy,in light of the demonstration.

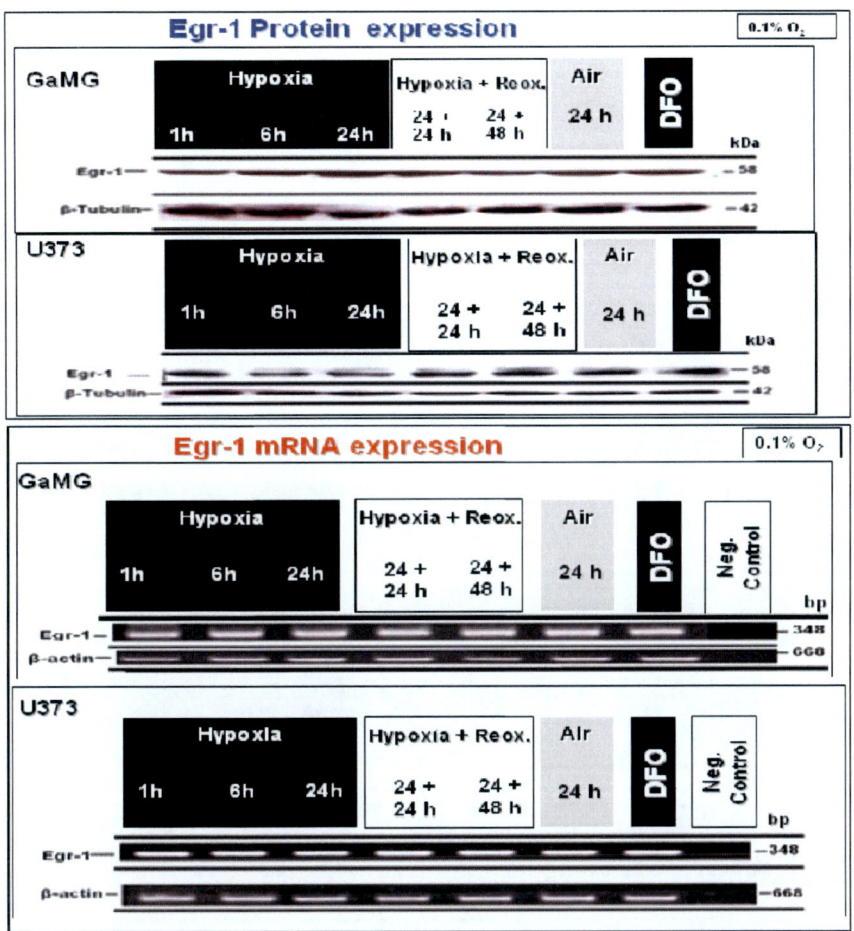

Figure 3. Egr-1 gene expression in response to hypoxia in human brain tumor cells.
Upper panels: Regulation of Epidermal growth factor-1 (Egr-1) expression at the protein level
in vitro in human GBM cell lines exposed to hypoxic (0.1% O2) conditions for 1h, 6 h or
24h. Protein laysates subjected to 24h under normoxia or to treatment with desferroxamine
(100μM) served as negative and positive control, respectively.
Lower panels: Examination of hypoxia-inducible regulation of epidermal growth factor-1 (Egr-1) gene expression at the mRNA level (semiquantitative RT-PCR) *in vitro* in human GBM cell lines (hypoxia time course experiments at O_2 concentration of 0.1%. GBM cell lines exposed to hypoxic (0.1% O_2) conditions for 1h, 6 h or 24h. Total mRNA from cells subjected to 24h under normoxia or to treatment with desferroxamine (100μM) served as negative and positive control, respectively. RT-PCR did not reveal any regulatory event under different oxygenation, hypoxia and reoxygenation conditions. Bar graphs show band intensities after densitometric evaluation and normalisation to β-actin expression as it is known from previous experiments. Representativeexperiment out of three.

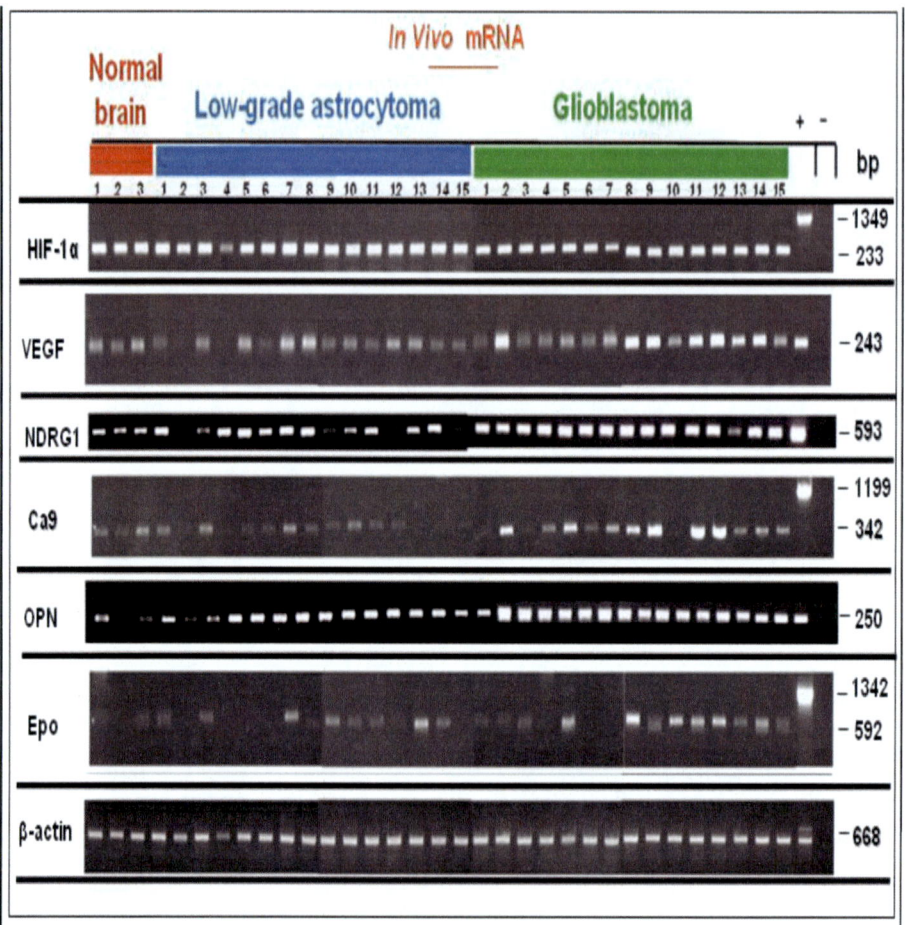

Figure 4. Hypoxia induced genes mRNA expression in correlation with grading of human brain tumors.
Hypoxia induced genes mRNA expression in vivo in human GBM and LGA tumor samples. Expression of carbonic anhydrase IX (CA IX), osteopontin (OPN), erythropoietin (Epo), vascular endothelial growth factor (VEGF), N Myc- downregulated gene 1 NDRG1 and hypoxia-inducible factor-1a (HIF-1a) protein in a representative semi quatitative human tumor samples of low-grade astrocytoma (LGA) and glioblastoma (GBM). GaMG protein lysates subjected to 24 h under normoxia or to treatment with desferroxamine (100 μM) served as negative and positive control, respectively.

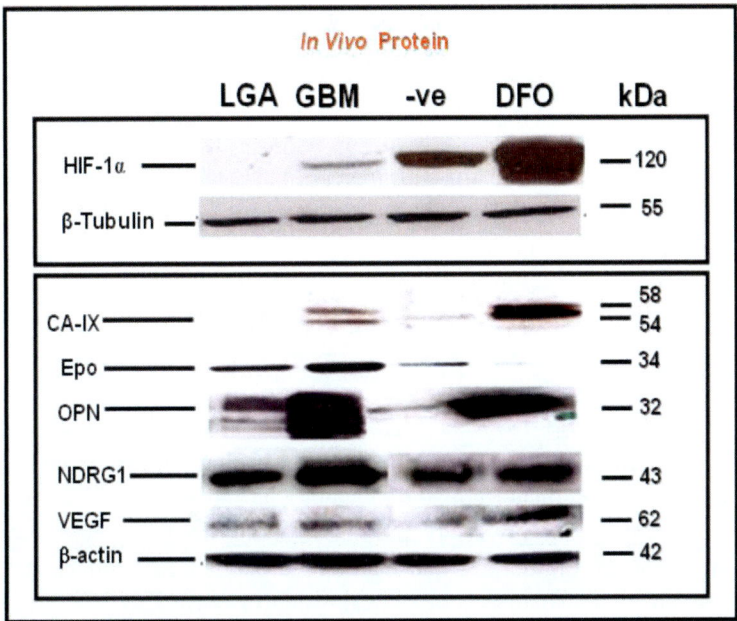

Figure 5. Hypoxia induced genes protein expression in correlation with grading of human brain tumors. Expression of carbonic anhydrase IX (CA IX), osteopontin (OPN), erythropoietin (Epo), vascular endothelial growth factor (VEGF), N Myc- downregulated gene 1 NDRG1 and hypoxia-inducible factor-1a (HIF-1a) protein in a Western blot of representative human tumor samples of low-grade astrocytoma (LGA) and glioblastoma (GBM). GaMG protein lysates subjected to 24 h under normoxia or to treatment with desferroxamine (100 μM) served as negative and positive control, respectively.

Furthermore, the mRNA expression levels for the other individual genes in patient tumor samples (GBM and LGA) were determined for EPO, OPN, CA9, VEGF (Figure 4). OPN mRNA was found uniformly upregulated in GBM, compared to a normal brain and to LGA. CA9 exhibited a strong upregulation in GBM in about half of the tumor specimens (7/15 greater induction than the highest seen in (LGA); EPO was overexpressed in GBM as compared to normal brain, but not significantly so compared to LGA, due to rather strong expression in some LGA samples. VEGF expression was over 2-fold of normal brain expression in 2/15 (LGA) and 7/15 (GBM), resulting in a significant overexpression in GBM vs. LGA,while on the experiments examiningthe tumor on protein levels (Figure 5) we found that all these genes where clearly upregulated, as well as HIF-1α.

Hypoxia Induced NDRG1 Expression in Human Brain Tumor Cells

Hypoxia is a factor that playsan important role within the solid tumor microenvironment. It significantly influences the behaviour of tumor cells via activation of genes encoding proteins involved in adaptation to hypoxic stress. Also, it plays an important role in tumor progression. It select cells with enhanced glycolytic activity, causing production of large amounts of lactic acid, one of the most common features of tumor cells (Warburg effect). As an important phenomenon, it attracts a lot of attention owing to its significant correlation with the tumor progression, treatment result and the overall prognosis of disease [Brown JM2000, Kallio PJ 1997]. Tumor tissue growth requires a sufficient supply of oxygen and nutrients. However, proliferating tumor cells quickly overgrow the diffusion distance of the oxygen from the nearest blood vessel (100 -150 µm), leading to a tumor vasculature that is highly irregular and tortuous, with arteriovenous shunts, blind ends, and incomplete endothelial linings. This has an effect on blood flow, which is less efficient than in normal tissues [Richard DE 1999, Guppy M 2002]. Tumor expansion is characterized by rapid growth of cancer cells when tumors establish themselves in host tissues of organs. Rapid growth of the tumor accompanies alterations in the cancer cell microenvironment, caused by an inability of local vasculature to supply enough oxygen and nutrients to the rapidly dividing tumor cells [Qu X 2002]. This makes hypoxia one common feature of solid tumors [Angst E 2006].NDRG1 is a member of the N – myc downregulated gene (NDRG) family.*NDRG1* (also known as *Drg1, RTP, Rit42,PROXY-1*, and *Cap43*) was identified as a gene up-regulated duringcellular differentiation [van Belzen N 1997, Bandyopadhyay S 1999Piquemal D 2003].

Itis induced by hypoxia [Salnikow K 2003, Greijer AE 2005]. HIF-1 as a transcription factor plays a major role in the regulation of hypoxia-responsive genes [Lachat P 2002, Chen B 2006,Sibold S 2007] and is also involved in the transcriptional regulation of the *NDRG1* gene [Han YH 2007, Cangul H 2004], together with other transcription factors. In this relation it is of interest to investigate the expression of NDRG1 protein in human cancer [Stein S 2004]. This gene is necessary for p53-mediated apoptosis and regulated by PTEN (phosphatase and tensin homologue). In several cancers, it was suggested to be a tumor suppressor gene [Ando T2006].

In several experimental analyses for the activity of hypoxia induced genes, in two collected groups of tumor specimens from human patients, [one group consisted of patients with low-grade astrocytoma while the other one was comprised of patients suffering from glioblastoma]we found that *in vivo*, mRNA expression of HIF-1· was similar in tumor specimens from patients with low-grade astrocytoma (LGA) or glioblastoma (GBM) with a group of normal brain tissue samples as a control group(Figure 4). A tumor-grade association of NDRG1 mRNA expression was exhibited *in vivo*. No increase in NDRG1 expression was shown in low-grade astrocytoma,while a two-fold increase in NDRG1 expression was shown in 10/15 patients withglioblastoma. NDRG1 protein (Figure 5) is also higher expressed in GBM when compared to (LGA).

Chapter 4

Regulation via other Hypoxia Gene Regulators

Early growth response factor 1 (Egr-1) is a transcription factor that triggers transcription of downstream genes within 15 - 30 min of various stimulations [Akutagawa O 2008].

In several publications [Ellen T 2008, Zhang P 2007] it has been shown that the transcription factor Egr-1 is regulated via hypoxia, and hypothesized to be responsible for the hypoxia induced regulation of N-myc downregulated gene 1 (NDRG1) in human tumours cells. HIF-1α is a transcription factor consisting of two HIF-1 basic-helix-loop-helix proteins containing a PAS domain subunits [1995 Wang GL]

In normoxia, the von Hippel-Lindau tumour suppressor (pVHL),which is the recognition component of an E3 ubiquitin ligasecomplex, targets HIF-1α[Iwai K 1999, Lisztwan J 1999].This leads to its ubiquitylation and consequent proteasomal degradation [Lisztwan J 1999,Cockman ME 2000, Kamura T 2000,Ohh M 2000, Tanimoto K2000].

HIF-1α is responsible forthe regulation of several hypoxia-induced genes. These genes are expressed rapidly through specific promoter activation by binding to the so called 'hypoxia responsive element' (HRE) and by doing that,they mediate cell growth and angiogenesis [Dachs GU 1995, Melillo G1997]. Due to HRE positioning in the distal promoter region, HIF-1 acts as a transcriptional enhancer, as hypothesized in a similar gene expression model [Wang GL1995]. In hypoxia, the α/β heterodimer binds to the core pentanucleotide sequence (RCGTG) in the hypoxia response elements (HREs) of target genes. HIF-β subunits are non-oxygen responsive nuclear proteins that also have other roles in transcription [2000 Gu YZ].

In contrast, the HIF-α subunits are highly inducible by hypoxia responsive element (HRE) bound by nuclear HIF-1α in human glioblastoma cells *in vitro* under different oxygenation conditions. Also, the clear enhanced binding of nuclear extracts from glioblastoma cell samples exposed to extreme hypoxic conditions confirms the HIF-1 western results [Said HM 2008, Said HM 2009] and the NDRG1 regulation by HIF-1 α [Said H 2006, Said HM 2007].

Our findings demonstrate that EGR-1 is not up-regulated in response to the extreme hypoxic (0.1 %O_2) or even reoxygenative conditions after hypoxia in human glioblastoma. Therefore, HIF-1α is still one of the most promising targets for new therapeutic strategies in cancer research, especially therapeutic modulation of the adaptive hypoxic response. At the same time we believe, based on the data that resulted from the research work of the different research groups worldwide, that Egr-1 could not be a new target for therapeutic modulation of the adaptive hypoxic response, at least not in human glioblastoma.

Conclusions

Hypoxia significantly influences the behaviour of human tumor cells via activation of genes involved in the adaptation to hypoxic stress, representing an important indicator of cancer prognosis and is associated with aggressive growth, metastasis, and poor response to treatment and with malignant progression

In malignant glioma therapy, the main aim, as with all cancers, is to either eradicate the tumor or convert it into a controlled, quiescent chronic disease. Angiogenesis and hypoxia induced, HIF-1α regulated gene inhibition remain the main parts of therapeutic approaches in human oncology. It is well known that cancer cell metabolism can be perturbed specifically at the level of glycolysis leading to interesting therapeutic activities in cancer that can be displayed. This functional characteristic was applied as proof-of-concept tools, demonstrating that CA IX and also its main regulator HIF-1α represent interesting targets for anticancer drug development. The differences between different types of brain tumors and brain tumor cell lines regarding the responseof the hypoxia induced genes CA9, EPO, OPN, VEGF,and NDRG1 to hypoxia and reoxygenation which was observed here can, in part, be explained by the genetic background of the cell lines investigated which may indirectly influence their degree of expression.Glycolytic inhibitors, when added in controlled doeses under hypoxia, lead to a reduced accumulation of HIF-1α and can function as indirect inhibitors of hypoxia genes like OPN, EPO, VEGF, NDRG1 and CA9 besides other direct gene inhibitors like sulphonamide derivates or Chetomin and other tools likes siRNA.This rendersthem optimal tools for the development of optimized therapyforhuman cancer treatment, especially in human brain cancer treatment.

Acknowledgments

The authors would like to thank Astrid Katzer, Stefanie Gerngras and Siglinde Kühnelfor technical assistance during the different stages of this analysis. The author would like to thank The University of Würzburg, Medical Faculty, Department of Radiation Oncology the Deutsche Forschung-sgemeinschaft (VO 871/2-3) to DV and by IZKF Würzburg (B25) to CH and GHV for financing this research. Also, the authors would like to acknowledge the efforts & contributions made by the different research groups in tumor hypoxia hypoxia signalling and regulation.

References

Acs, G., Zhang, P. J., Rebbeck, T. R., Acs, P. & Verma, A. (2002). Immunohistochemical expression of erythropoietin and erythropoietin receptor in breast carcinoma. *Cancer,* 95, 969–981.

Acs, G., Zhang, P. J., McGrath, C. M., Acs, P., McBroom, J., Mohyeldin, A., Liu, S., Lu, H. & Verma, A. (2003). Hypoxia-inducible erythropoietin signaling in squamous dysplasia and squamous cell carcinoma of the uterine cervix and its potential role in cervical carcinogenesis and tumor progression. *Am J Pathol,* 162, 1789–1806.

Aebersold, D. M., Burri, P., Beer, K. T. et al. (2001). Expression of hypoxia-inducible factor- 1α: a novel predictive and prognostic parameter in the radiotherapy of oropharyngeal cancer. *Cancer Res,* 61, 2911–6.

Ahmad IM, Aykin - Burns N, Sim JE, et al, (2005) Mitochondrial O2 and H2O2 mediate glucose deprivation-induced cytotoxicity and oxidative stress in human cancer*cells. J Biol Chem;* 280: 4254 - 63.

Akutagawa, O., Nishi, H., Kyo, S., Terauchi, F., Yamazawa, K., Higuma, C., Inoue, M. & Isaka, K. (2008) Early growth response-1 mediates downregulation of telomerase in cervical cancer. *Cancer Sci.,* 99(7), 1401-6.

Alabovskii, V. V,Khamburov, V. V & Vinokurov, A. A, (2004). Some mechanisms of adenosine protective effect in the "calcium paradox". *Ross. Fiziol. Zh. Im. I. M.* Sechenova, 90(7), 889-901.

Ando, T., Ishiguro, H., Kimura, M. et al. (2006). Decreased expression of NDRG1 is correlated with tumor progression and poor prognosis in patients with esophageal squamous cell carcinoma. *Dis. Esophagus,* 19, 454–458.

Angst, E., Sibold, S., Tiffon, C. et al. (2006). Cellular differentiation determines the expression of the Hypoxia - inducible protein NDRG1 in pancreatic cancer. *Br. J. Cancer,* 95, 307-13.

Arcasoy, M. O., Amin, K., Karayal, A. F., Chou, S. C., Raleigh, J. A., Varia, M. A. & Haroon, Z. A. (2002). Functional significance of erythropoietin receptor expression in breast cancer. *Lab Invest,* 82, 911–918.

Bache, M., Reddemann, R., Said, H. M. et. al. (2006). Immunohistochemical detection of osteopontin in advanced head-and-neck cancer: prognostic role and correlation with oxygen electrode measurements, hypoxia-inducible-factor-1alpha-related markers, and hemoglobin levels. *Int J Radiat Oncol Biol Phys,* 66(5),1481-7.

Bandyopadhyay, S., Pai, S. K., Gross, S. C., et al. (2003). The Drg-1 gene suppresses tumor metastasis in prostate cancer. *Cancer Res,* 63, 1731–6.

Batra, S., Perelman, N., Luck, L. R., Shimada, H. & Malik, P. (2003). Pediatric tumor cells express erythropoietin and a functional erythropoietin receptor that promotes angiogenesis and tumor cell survival. *Lab Invest,* 83, 1477–1487.

Birner, P., Schindl, M., Obermair, A. et al. (2001). Expression of hypoxia-inducible factor-1α in epithelial ovarian tumors: its impact on prognosis and on response to chemotherapy. *Clin Cancer Res,* 7, 1661–8.

Bos, R., van der Groep, P., Greijer, A. E. et al. (2003). Levels of hypoxia-inducible factor- 1α independently predict prognosis in patients with lymph node negative breast carcinoma. *Cancer,* 97, 1573–81.

Brizel, D. M., Schroeder, T., Scher, R. L. et al. (2001). Elevated tumor lactate concentrations predict for an increased risk of metastases in head-and-neck cancer. *Int J Radiat Oncol Biol Phys,* 51, 349–53.

Brizel, D. M., Scully, S. P., Harrelson, J. M., Layfield, L. J., Bean, J. M., Prosnitz, L. R. & Dewhirst, M. W. (1996). Tumor oxygenation predicts for the likelihood of distant metastases in human soft tissue sarcoma. *Cancer Res.,* 56, 941-943.

Brown, J. M. (2000). Exploiting the hypoxic cancer cell: mechanisms and therapeutic strategies. *Mol. Med. Today,* 6, 157–162.

Brown JM, QT Le, (2002), Tumor hypoxia is important in radiotherapy, but how should we measure it?, *Int. J Radiat. Oncol. Biol. Phys* 54 (5): 1299 -1301

Brown LM, Cowen RL, Debray C, et al. (2006), Reversing hypoxic cell chemoresistance in vitro using genetic and small molecule approaches targeting hypoxia inducible factor. *Mol Pharmacol;* 69: 411-8.

Bunn, H. F. & Poyton, R. O. (1996). Oxygen sensing and molecular adaptation to Hypoxia. *Physiol. Rev.,* 76,839–885.

Bunworasate, U., Amouk, H., Mindeman, H., O'Loughlin, K. L., Sait, S. N. J., Barcos, M., Stewart, C. C. & Baer, M. R. (2001). Erythropoietindependent transformation of myelodysplastic syndrome to acute monoblastic leukemia. *Blood,* 98, 3492–3494.

Burk, D., Woods, M. & Hunter, J. (1967). On the significance of glycolysis for cancer growth, with special reference to Morris rat hepatomas. *J Natl Cancer Inst,* 38, 839–63.

Beasley NJ, Wykoff CC, Watson PH, Leek R, Turley H, Gatter K, Pastorek J Cox GJ, Ratcliffe P, Harris AL (2001), Carbonic anhydrase IX, an endogenous hypoxia marker, expression in head and neck squamous cell carcinoma and its relationship to hypoxia, necrosis, and microvessel density, *Cancer Res.* 61(13):5262-7.

Bui MH, Seligson D, Han KR, Pantuck AJ, Dorey FJ, Huang Y, Horvath S,(2003), Carbonic anhydrase IX is an independent predictor of survival inadvanced renal clear cell carcinoma: implications for prognosis and therapy. *Clin. Cancer Res.* 9(2):802-11.

Bishop JM (1987), The molecular genetics of cancer, *Science* 235(4786):305-11.

Colpaert, C. G. et al. (2003).Cutaneous breast cancer deposits show distinct growth patterns with different degrees of angiogenesis, hypoxia and fibrin deposition. *Histopathology* 42,530–540

Cangul, H. (2004). Hypoxia upregulates the expression of the NDRG1 gene leading to its overexpression in various human cancers. *BMC Genet,* 5, 27.

Chen, B., Nelson, D. M., Sadovsky, Y. (2006). N - Myc downregulated gene 1 (Ndrg1) modulates the response of term human trophoblasts to hypoxic injury. *J. Biol. Chem,* 281, 2764 – 2772.

Chrastina A, Pastoreková S, Pastorek J.(2003), Immunotargeting of human cervical carcinoma xenograft expressing CA IX tumor-associated antigen by 125I-labeled M75 monoclonal antibody. *Neoplasma.* 50 (1): 13 - 21.

Chan, D. A. & Giaccia, A. J. (2007), Hypoxia, gene expression, and metastasis. *Cancer Metastasis Rev.* 26, 333–339

Chomczynski, P. & Sacchi, V. (1987). Single-step method of RNA isolation by acidic guanidinium thiocyanate-phenol-chloroform extraction. *Anal Biochem,* 162, 156-9.

Clarke, D. D. & Sokoloff, L. (1999). Circulation and energy metabolism in the brain. In G. J. Siegel, B. W. Agranoff, R. W. Albers, S. K. Fisher, & M. D. Uhler (Eds.) *Basic Neurochemistry* (6th edition, pp. 637-669). New York: Lippincott-Raven.

Cockman, M. E., Masson, N., Mole, D. R., Jaakkola, P., Chang, G. W., Clifford, S. C., Maher, E. R., Pugh, C. W., Ratcliffe, P. J. & Maxwell, P. H. (2000). Hypoxia inducible factor-alpha binding and ubiquitylation by the von Hippel-

Lindau tumor suppressor protein. *Journal -of Biological Chemistry,* 275, 25733–25741.

Coles, N. W. & Johnstone, R. M. (1962). *Biochem J,* 83, 284–291.

Dachs, G. U., Patterson, A. V., Firth, J. D., Ratcliffe, P. J., Townsend, K. M., Stratford, I. J. & Harris, A. L. (1997). Targeting gene expression to hypoxic tumor cells. *Nat Med.,* 3(5), 515-20.

Davis GE, Camarillo CW (1995), Regulation of endothelial cell morphogenesis by integrins, mechanical forces, and matrix guidance pathways. *Exp Cell Res* 216: 113–123

- Diaz-Gonzalez, J. A., Russell, J., Rouzaut, A., Gil-Bazo, I. & Montuenga, L. (2005). Targeting hypoxia and angiogenesis through HIF-1α inhibition. *Cancer Biol. Ther.* 4, 1055–1062

Dillard, D. G., Venkatraman, G., Cohen, C., Delgaudio, J., Gal, A. A. & Mattox, D. E. (2001). Immunolocalization of erythropoietin and erythropoietin receptor in vestibular schwannoma. *Acta Otolaryngol,* 121, 149–152.

Ebert, B. L., Firth, J. D. & Ratcliffe, P. J. (1995). Hypoxia and Mitochondrial Inhibitors Regulate Expression of Glucose Transporter-1 via Distinct Cis-acting Sequences. *J. Biol. Chem.,* 270, 29083 –29089.

Ellen, T., Ke, Q., Zhang, P. & Costa, M. (2008). NDRG1, a Growth and Cancer Related Gene: Regulation of Gene Expression and Function in Normal and Disease States. *Carcinogenesis,* 29(1), 2-8.

Eliceiri BP, Paul R, Schwartzberg PL, Hood JD, Leng J, Cheresh DA (1999). Selective requirement for Src kinases during VEGF-induced angiogenesis and vascular permeability. *Mol. Cell* 4, 915–924

Erler JT, Bennewith KL, Nicolau M, Dornhöfer N, Kong C, Le QT, Chi JT, Jeffrey SS, Giaccia AJ.(2006).Lysyl oxidase is essential for hypoxiainduced metastasis. *Nature* 440, 1222–1226

Evans, S. M. & Koch, C. J. (2003). Prognostic significance of tumor oxygenation in humans. *Cancer Lett,* 195, 1–16.

Evans SM, Hahn S, Pook DR, Jenkins WT, Chalian AA, Zhang P, Stevens C, Weber R, Weinstein G, Benjamin I, Mirza N, Morgan M, Rubin S, McKennaWG, Lord EM, Koch CJ, (2000), Detection of hypoxia in human squamous cell carcinoma by EF5 binding, *Cancer Res.* 60: 2018–2024.

Fandrey, J. (1995). Hypoxia-inducible gene expression, *Respir. Physiol.,* 101, 1–10.

Faulhammer F, Konrad G, Brankatschk B, Tahirovic S, Knödler A, Mayinger P (2005), Cell growth–dependent coordination of lipid signaling and glycosylation is mediated byinteractions between Sac1p and D pm1p. *J Cell Biol.* 168: 185-91.

Folkman J (1971): Tumor angiogenesis: therapeutic implications. *N Engl J Med* 285: 1182–1186

Floridi, A., Paggi, M. G. & Fanciulli, M. (1989). Modulation of glycolysis in neuroepithelial tumors. *J Neurosurg Sci,* 33, 55-64.

Galarraga, J., Loreck, D. J., Graham, J. F., DeLaPaz, R. L., Smith, B. H., Hallgren, D. & Cummins, C. J. (1986). Glucose metabolism in human gliomas: correspondence of in situ and in vitro metabolic rates and altered energy metabolism. *Metab Brain Dis,* 1, 279-291.

Geng L, Donnelly E, McMahon G, Lin PC, Sierra- Rivera E, Oshinka H, Hallahan DE (2001), Inhibition of vascular endothelial growth factor receptor signalling leads to reversal of tumor resistance to radiotherapy. *Cancer Res* 61:2413-2419

Giatromanolaki, A., Koukourakis, M., Sivridis, E. et al. (2001) Relation of hypoxia inducible factor 1α and 2α in operable non-small celllung cancer to angiogenic molecular profile of tumors and survival. *Br J Cancer,* 85, 881–90.

Galarraga J, Loreck DJ, Graham JF, DeLaPaz RL, Smith BH, Hallgren D, Cummins CJ (1986) Glucose metabolism in human gliomas: correspondence of in situ and in vitro metabolic rates and altered energy metabolism. *Metab Brain Dis*.1: 279 - 91.

Greene, A. E., Todorova, M. T. & Seyfried, T. N. (2003). Perspectives on the metabolic management of epilepsy through dietary reduction ofglucose and elevation of ketone bodies. *J Neurochem,* 86, 529-537.

Greenspan, R. J. (2001). The flexible genome. Nat Rev Genet, 2, 383-387.

Greijer, A. E., van der Groep, P., Kemming, D. et al. (2005). Up-regulation of gene expression by hypoxia is mediated predominantly by hypoxia-inducible factor 1 (HIF-1*). J. Pathol,* 206, 291–304.

Griffiths JR, McSheehy PM, Robinson SP, Troy H, Chung YL, Leek RD, Williams KJ, Stratford IJ, Harris AL, Stubbs M. (2002), Metabolic changes detected by in vivo magnetic resonance studies of HEPA-1 wild-type tumors and tumors deficient in hypoxia-inducible factor-1beta (HIF-1beta): evidence of an anabolic role for the HIF-1 pathway. *Cancer Res;* 62: 688 – 95.

Grumann, T., Arab, A., Bode, C. et al. (2006). Reoxygenation of human coronary smooth muscle cells suppresses HIF-1alpha gene expression and augments radiation-induced growth delay and apoptosis. *Strahlenther Onkol,* 182, 16–21.

Gu, Y. Z., Hogenesch, J. B. & Bradfield, C. A. (2000). The PAS superfamily: sensors of Environmental and developmental signals. *Ann. Rev Pharmacol Toxicol*, 40, 519–561.

Guppy, M. (2002) The hypoxic core: a possible answer to the cancer paradox. *Biochem. Biophys. Res. Commun,* 299, 676 – 80.

Hagen, T., Taylor, C.T., Lam, F. et al. (2003). Redistribution of intracellular oxygen in hypoxia by nitric oxide: effect on HIF1alpha. *Science,* 302, 1975–8.

Han, Y. H., Xia, L., Song, L. P. et al. (2007). Comparative proteomic analysis of hypoxia-treated and untreated human leukemic U937 cells. *Proteomics,* 6, 3262–3274.

Harris, R. A., Bowker-Kinley, M. M., Hyang, B. & Wu, P. (2002). Regulation of the activity of the pyruvate dehydrogenase complex. *Adv Enzyme Regul,* 42, 249–259.

Haroon, Z. A., Raleigh, J. A., Greenberg, C. S. & Dewhirst, M. W, (2000),Early wound healing exhibits cytokine surge without evidence of hypoxia. *Ann. Surg.* 231, 137–147

Holbrook JJ, Liljas A, Steindel SJ, Rossman MG (1975), Lactate dehydrogenase. In: Boyer PD (eds) *The enzymes.* Volume 11, 3rd ed. Academic Press, NY, pp 191–292

Hagen T,Taylor CT, Lam F, et al. (2003), Redistribution of intracellular oxygen in hypoxia by nitric oxide: effect on HIF1alpha.*Science* 302:1975 - 8.

Harris RA, Bowker - Kinley MM, Hyang B, Wu P (2002), Regulation of theactivity of the pyruvate dehydrogenase complex. *Adv Enzyme Regul.,* 42 :249– 59.

Holbrook JJ, Liljas A, Steindel SJ, Rossman MG, (1975), Lactate dehydrogenase. In: *Boyer PD, editor. Theenzymes.* Volume 11, 3rd ed. NY: Academic Press;.p.191 – 292

Hedeskov, C. J. (1968). Biochem J, 110, 373–380.

Heidenreich, S., Rahn, K. H. & Zidek, W. (1991). Direct vasopressor effect of recombinant human erythropoietin on renal resistance vessels. *Kidney Int,* 39, 259–265.

Hengartner, M. O. & Horvitz, H. R. (1994). C. elegans cell survival gene ced-9 encodes a functional homolog of the mammalian proto-oncogene bcl-2. *Cell,* 76, 665–676.

Hilvo, M., Rafajova, M., Pastorekova, S., Pastorek, J. & Parkkila, S. (2004). Expression of carbonic anhydrase IX in mouse tissues. *J Histochem. Cytochem.;* 52, 1313–22.

Höckel, M., Knoop, C., Schlenger, K., et al. (1993). Intratumoral pO2 predicts survival in advanced cancer of the uterine cervix. *Radiother Oncol,* 26, 45–50.

Höckel, M., Schlenger, K., Aral, B. et al. (1996). Association between tumor hypoxia and malignant progression inadvanced cancer of the uterine cervix. *Cancer Res,* 56, 4509–15.

Höckel, M. & Vaupel, P. (2001). Tumor hypoxia: definitions and current clinical, biologic, and molecular aspects. *J Natl Cancer Inst,* 93, 266–76.

Holbrook, J. J.,Liljas, A., Steindel, S. J. & Rossman, M. G. (1975). Lactate dehydrogenase. In P. D. Boyer (Eds.) *The enzymes* (3rd edition, pp. 191–292). NY: Academic Press.

Huang, L. E., Arany, Z., Livingston, D. M. & Bunn, H. F. (1996). Activation of Hypoxia-inducible Transcription Factor Depends Primarily upon Redox-sensitiveStabilization of Its α Subunit, *J. Biol. Chem.,* 271, 32253-32259.

Isidoro, A., Casado, E., Redondo, A. et al. (2005) Breast carcinomas fulfill the Warburg hypothesis and provide metabolic markers of cancer prognosis. *Carcinogenesis,* 26, 2095–104.

Ivanov, S., Liao, S. Y., Ivanova, A., Danilkovich-Miagkova, A., Tarasova, N., Weirich, G., Merrill, M. J., Proescholdt, M. A., Oldfield, E. H., Lee, J., Zavada, J., Waheed, A., Sly, W., Lerman, M. I., Stanbridge, E.J. (2001). Expression of hypoxia-inducible cell-surface transmembrane carbonic anhydrases in human cancer. *Am. J. Pathol.,* 158, 905 - 919.

Iwai, K., Yamanaka, K., Kamura, T., et al. (1999). Identification of the von Hippellindau tumor-suppressor protein as part of an active E3 ubiquitin ligase complex. *PNAS,* 96, 12436–12441.

Ingber DE, Prusty D, Sun Z, Betensky H, Wang N (1995), Cell shape, cytoskeletal mechanics, and cell cycle control in angiogenesis. *J Biomech* 28: 1471–1484

Isidoro A, Casado E, Redondo A, Acebo P, Espinosa E, Alonso AM, Cejas P, Hardisson D, Fresno Vara JA, Belda-Iniesta C, González-Barón M, Cuezva JM. (2005) Breast carcinomas fulfill the Warburg hypothesis and provide metabolic markers of cancer prognosis. *Carcinogenesis* 26: 2095–2104.

Kacser, H., Burns, J. A. (1981). The molecular basis of dominance. *Genetics,* 97, 639 -666.

Jang YC, Arumugam S, Gibran NS, Isik FF (1999), Role of alpha V integrins and angiogenesis during wound repair. *Wound Repair Regen.* 7: 375–380

Kallio, P. J., Pongratz, I., Gradin, K., McGuire, J. & Poellinger, L. (1997). Activation of hypoxia-inducible factor 1α: Posttranscriptional regulation and conformational change by recruitment of the Arnt transcription factor. *PNAS,* 94, 5667-5672.

Kaanders JH, Wijffels KI, Marres HA, Ljungkvist AS, Pop LA, van den HoogenFJ, de Wilde PC, Bussink J, Raleigh JA, van der Kogel AJ (2002),

Pimonidazole binding and tumor vascularity predict for treatment outcome inhead and neck cancer, *Cancer Res.* 62: 7066–7074

Kamura, T., Sato, S., Iwai, K., Czyzyk-Krzeska, M., Conaway, R. C. & Conaway, J. W. (2000). Activation of HIF1alpha ubiquitination by a reconstituted von Hippel-Lindau (VHL) tumor suppressor complex. *PNAS,* 97, 10430–10435.

Kim, S. J., Rabbani, Z. N., Vollmer, R. T., Schreiber, E. G., Oosterwijk, E., Dewhirst, M. W., Vujaskovic, Z. & Kelley, M. J. (2004). Carbonic anhydrase IX in early-stage non small cell lung cancer. Clin Cancer Res, 10, 7925-33.

Knisely, J., Rockwell, S. & Knisely, J. (2002). The importance of hypoxia in the brain tumor, *Neuroimaging Clin. N. Amer.,* 12, S25–S35.

Koukourakis, M. I., Bentzen, S. M., Giatromanolaki, A., Wilson, G. D., Daley, F. M., Saunders, M. I., Dische, S., Sivridis, E., Harris, A. L. (2006). Endogenous markers of two separate hypoxia response pathways (hypoxia inducible factor 2 alpha and carbonic anhydrase 9) are associated with radiotherapy failure in head and neck cancer patients recruited in the CHART randomized trial. *J. Clin. Oncol.,* 24(5), 727-35.

Kurimoto M, Endo S, Hirashima Y, Nishijima M, Takaku A (2001). Elevated plasma basic fibroblast growth factor in brain tumor patients. *Neurol Med Chir* (Tokyo) 36: 865–868; 869, 1996

Kuroi, K. & Toi, M. Circulating angiogenesis regulators in cancer patients. *Int. J. Biol. Markers* 16, 5 -26

Kung AL, Zabludoff SD, France DS, Friedmann SJ, Tanner EA, Vieira A, Cornell-Kennon S, Lee J, Wang B, Wang J, Memmert K, Naegeli HU, Petersen F, Eck MJ, Bair KW, Wood AW, Livingston DM: Small molecule blockade of transcriptional coactivation of the hypoxia-inducible factor pathway. *Cancer Cell* 2004, 6: 33 – 43

Kwon SJ, Lee YJ. Effect of low glutamine/glucose on hypoxia-induced elevation of hypoxia-inducible factor-1alpha in human pancreatic cancer MiaPaCa-2 and human prostatic cancer DU-145 cells. *Clin Cancer Res* 2005; 11: 4694 - 4700.

Lachat, P., Shaw, P., Gebhard, S., et al. (2002). Expression of NDRG1, a differentiation-related gene, in human tissues. *Histochem Cell. Biol,* 118, 399–408.

Lee, J. W., Bae, S. H., Jeong, J. W. et al. (2004). Hypoxia-inducible factor (HIF-1) alpha: its protein stability and biological functions. *Exp Mol Med,* 36, 1–12.

Laderoute, K. R. et al. (2000) Opposing effects of hypoxia on expression of the angiogenic inhibitor thrombospondin 1 and the angiogenic inducer vascular endothelial growth factor. *Clin. Cancer Res.* 6, 2941–2950

Lisztwan, J., Imbert, G., Wirbelauer, C., Gstaiger, M. & Krek, W. (1999). The von Hippel-Lindau tumor suppressor protein is a component of an E3 ubiquitin-protein ligase activity. *Genes & Development,* 13, 1822–1833.

Linderholm B, Tavelin B, Grankvist K, Henriksson R (1999),Does vascular endothelial growth factor (VEGF) predict local relapse and survival in radiotherapy-treated node-negative breast cancer? *Br J Cancer* 81:727-732

Lin X, Zhang F, Bradbury CM, et al. (2003) 2-Deoxy-D-glucose-induced cytotoxicity and radiosensitization in tumor cells is mediated via disruptions in thiol metabolism. *Cancer Res;*63:3413–17.

Liu L, Cash TP, Jones RG, et al. (2006), Hypoxia-induced energy stress regulates mRNA translation and cell growth. *Mol Cell.*21:521–31.

Liu L, Simon C. (2004), Regulation of transcription and translation by hypoxia. *Cancer Biol Ther;3*: 492 - 7.

Lu H, Forbes RA, Verma A. (2002), Hypoxia-inducible factor 1 activation by aerobic glycolysis implicates the Warburg effect in carcinogenesis. *J Biol Chem.,* 277: 23111-5.

Madan, A., Lin, C., Wang, Z. & Curtin, P. T. (2003). Autocrine stimulation by erythropoietin in transgenic mice results in erythroid proliferation without neoplastic transformation. *Blood Cells Mol Dis,* 30, 82–89.

Mangiardi, J. R. & Yodice, P. (1990). Metabolism of the malignant astrocytoma. *Neurosurgery,* 26, 1-19.

Mayer, R., Hamilton-Farrell, M. R., van der Kleij, A. J. et al. (2005). Hyperbaric oxygen and radiotherapy. *Strahlenther Onkol,* 181, 113–23.

Maher JC, Krishan A, Lampidis TJ. (2004) Greater cell cycle inhibition and cytotoxicity induced by 2 deoxy-D-glucose in tumor cells treated under hypoxic vs aerobic conditions. *Cancer Chemother Pharmacol;*53: 116 - 22.

Mateo J, Garcia-Lecea M, Cadenas S,et al. (2003),Regulation of hypoxia-inducible factor-1alpha by nitric oxide through mitochondria-dependent and -independent pathways. *Biochem J;*376:537-44.

Moeller BJ, Cao Y, Li CY, Dewhirst MW (2004),Radiation activates HIF-1 to regulate vascular radiosensitivity in tumors: role of reoxygenation, free radicals, and stress granules. *Cancer Cell,* 5: 429 – 441

Mazure, N. M., Chen, E. Y., Yeh, P., Laderoute, K. R. & Giaccia, A. J (1996). Oncogenic transformation and hypoxia synergistically act to modulate vascular endothelial growth factor expression. *Cancer Res.* 56, 3436–3440

Macheda ML, Rogers S, Best JD, (2005), Molecular and cellular regulation of glucosetransporter (GLUT) proteins in cancer,*J. Cell. Physiol,* 202:654–62.

Melillo, G., Musso, T., Sica, A., Taylor, L. S., Cox, G. W. & Varesio, L. (1995). A hypoxia-responsive element mediates a novel pathway of activation of the inducible nitric oxide synthase promoter. *J Exp Med.,* 182(6), 1683-93.

Mies, G., Paschen, W., Ebhardt, G. & Hossmann, K. A. (1990). Relationship between of blood flow, glucose metabolism, protein synthesis, glucose and ATP content in experimentally-induced glioma (RG1 2.2) of rat brain. *J Neurooncol,* 9, 17-28.

Molineux, G. (2003). Biology of erythropoietin. In: G. Molineux, M. A. Foote, & S. G. Elliot (Eds.), *Erythropoietins and erythropoiesis.* (pp. 113-133). Basel: Birkhäuser.

Molls, M., Stadler, P., Becker, A., et al. (1998). Relevance of oxygen in radiation oncology. Mechanisms of action, correlation to low hemoglobin levels. *Strahlenther Onkol,* 174, Suppl: 13–6.

Nagamatsu, S., Nakamichi, Y., Inoue, N., Inoue, M., Nishino, H. & Sawa, H. (1996). Rat C6 glioma cell growth is related to glucose transport and metabolism. *Biochem J,* 319 (Pt 2), 477-482.

Nodin, C., Nilsson, M. & Blomstrand, F. (2005). Gap junction blockage limits intercellular spreading of astrocytic apoptosis induced by metabolic depression. *Journal of Neurochemistry,* 94(4), 1111-23.

Nordsmark, M., Bentzen, S. M., Rudat, V. et al. (2005). Prognostic value of tumor oxygenation in 397 head and neck tumors after primary radiation therapy. An international multi-center study. *Radiother Oncol,* 77, 18 – 24.

Ohh, M., Park, C. W., Ivan, M., Hoffman, M. A., Kim, T. Y., Huang, L. E., Pavletich, N., Chau, V. & Kaelin, W. G. (2000). Ubiquitination of hypoxia-inducible factor requires direct binding to the beta-domain of the von Hippel-Lindau protein. *Nature Cell Biology,* 2, 423–427.

Opavsky, R., Pastoreková, S., Zelník, V., Gibadulinová, A., Stanbridge, E. J., Závada, J., Kettmann, R. & Pastorek, J. (1996). Human MN/CA9 gene, a novel member of the carbonic anhydrase family: structure and exon to protein domain relationship. *Genomics,* 33, 480-487.

Oudard, S., Boitier, E., Miccoli, L., Rousset, S., Dutrillaux, B. & Poupon, M. F. (1997). Gliomas are driven by glycolysis: putative roles of hexokinase, oxidative phosphorylation and mitochondrial ultrastructure. *Anticancer Res,* 17, 1903-1911.

Overgaard, J. (1989). Sensitization of hypoxic tumour cells – clinical experience, *Int J. Radiat. Biol.,* 56, 801 – 11.

Olive PL, Aquino-Parsons C, MacPhail SH, Liao SY, Raleigh JA, Lerman MI, StanbridgeEJ, (2001),Carbonic anhydrase 9 as an endogenous marker forhypoxic cells in cervical cancer, *Cancer Res.* 61(24):8924-9.

Parkkila, S., Rajaniemi, H., Parkkila, A. K., Kivelä, J., Waheed, A., Pastoreková, S., Pastorek, J. & Sly, W. S. (2000). Carbonic anhydrase inhibitor suppresses invasion of renal cancer cells in vitro. *PNAS*, 5, 2220–2224.

Pastorek, J., Pastoreková, S., Callebaut, I., Mornon, J. P., Zelník, V., Opavsky, R., Zatóvicová, M., Liao, S., Portetelle, D., Stanbridge, E. J., Závada, J., Burny, A. & Kettmann, R. (1994). Cloning and characterization of MN, a human tumor- associated protein with a domain homologous to carbonic anhydrase and a putative helix-loop-helix DNA binding segment, *Oncogene*, 9, 2788-2888.

Pastoreková, S., Parkkila, S., Parkkila, A. K., Opavsky, R., Zelník, V., Saarnio, J. & Pastorek, J. (1997). Carbonic anhydrase IX, MN/CA IX: analysis of stomach complementary DNA sequence and expression in human and rat alimentary tracts. *Gastroenterology*, 112, 398-408.

Pastorekova, S. & Zavada, J. (2004). Carbonic anhydrase IX (CA IX) as a potential target for cancer therapy. *Cancer Ther*, 2, 245–62.

Pastorekova S, Casini A, Scozzafava A, Vullo D, Pastorek J, Supuran CT (2004). Carbonic anhydrase inhibitors: the first selective, membrane-impermeant inhibitors targeting the tumor-associated isozyme IX. *Bioorg Med. Chem Lett.* 23; 14(4): 869 - 73.

Prendergast G C, Cole M D (1989), Posttranscriptional regulation of cellular gene expression by the c-myc oncogene, *Mol. Cell Biol.* 9(1): 124–134.

Pfeiffer, T., Schuster, S. & Bonhoeffer, S. (2001). Cooperation and competition in the evolution of ATP-producing pathways. *Science*, 292, 504–7.

Piquemal, D., Joulia, D. & Commes, T. (1999). Transforming growth factor-ß1 is an autocrine mediator of U937 cell growth arrest and differentiation induced by vitamin D3 and retinoids. *Biochim. Biophys. Acta*, 1450, 364–73.

Plate KH, Breier G, Weich HA, RisauW (1992), Vascular endothelial growth factor is a potential tumour angiogenesis factor in human gliomas in vivo. *Nature* 359: 845–848

Qu, X. Zhai, Y. Wei, H. et al. (2002). Characterization and expression of three novel differentiation - related genes belong to the human NDRG gene family. *Mol Cell Biochem.*, 229, 35–44.

Reitzer, L. J., Wice, B. M. & Kennell, D. (1979). *J Biol Chem*, 254, 2669–2676.

Rhodes, C. G., Wise, R. J., Gibbs, J. M., Frackowiak, R. S., Hatazawa, J., Palmer, A. J., Thomas, D. G. & Jones, T. (1983). In vivo disturbance of the oxidative metabolism of glucose in human cerebral gliomas. *Ann Neurol*, 14, 614-626.

Richard, D. E., Berra, E. & Pouyssegur, J. (1999). Angiogenesis: how a tumor adapts to hypoxia. *Biochem. Biophys. Res. Commun.*, 266, 718-722.

Rohrer Bley, C., Ohlerth, S., Roos, M., et al. (2006). Influence of pretreatment polarographically measured oxygenation levels in spontaneous canine tumors treated with radiation therapy. *Strahlenther Onkol,* 182, 518 – 24.

Roslin, M., Henriksson, R., Bergstrom, P., Ungerstedt, U. & Bergenheim, A. T. (2003). Baseline levels of glucose metabolites, glutamate and glycerol in malignant glioma assessed by stereotactic microdialysis. *J Neurooncol,* 61, 151-160.

Rofstad, E. K. (2000).Microenvironment-induced cancer metastasis. *Int. J. Radiat. Biol.* 76, 589–605

Said, H. M., Katzer, A., Flentje, M. & Vordermark, D. (2005) Response of the plasma hypoxia marker osteopontin to in vitro hypoxia in human tumor cells. *Radiother Oncol.,* 76, 200-5.

Said, H. M., Katzer, A., Flentje, M. & Vordermark D. (2006). Response of the plasma hypoxia marker osteopontin to in vitro hypoxia in human tumor cells. *Radiother Oncol., Letter to the editor,* 78, 230–231.

Said, H. M., Polat, B., Hagemann, C. et al. (2008). Rapid detection of the hypoxia-regulated CA-IX and NDRG1 gene expression in different glioblastoma cells in vitro. *Oncology Rep.,* 20, 413-419.

Said, H. M., Staab, A., Hagemann, C. et. al. (2007). Distinct patterns of hypoxic expression of carbonic anhydrase IX (CA IX) in human malignant glioma cell lines. *J Neurooncol,* 81, 27-38.

Said, H. M., Stein, S., Hagemann, C. et al. (In Press). Oxygen-dependent regulation of NDRG1 in human glioblastoma cells in vitro and in vivo. *Oncology Rep.*

Said, H. M., Stein, S., Hagemann, C., Polat, B., Schömig, B., Staab, A., Theobald, M., Flentje, M., Vordermark, D. (2007). NDRG1 regulation as a response to an alternating hypoxic microenviroment invivoand in vitro in human brain tumors. *FEBS j,* 274 (s1), 281.

Said H, Stein S, Staab A, Katzer A, Flentje M, Vordermark D:NDRG1 is regulated in human glioblastoma in vitro as a consequence to the changing concentrations of the oxygen Microenviroment.*FEBS j*273 (s1):345, 2006

Sakata, K., Someya, M., Nagakura, H. et al. (2006). A clinical study of hypoxia using endogenous hypoxic markers and polarographic oxygen electrodes. *Strahlenther Onkol,* 182, 511–7.

Salnikow, K., Davidson, T., Zhang, Q. et al. (2003). The involvement of hypoxia-inducibletranscription factor-1-dependent pathway in nickel carcinogenesis. *Cancer Res,* 63, 3524–3530.

Schmaltz, C., Hardenbergh, P. H., Wells, A. & Fisher, D. E. (1998). Regulation of proliferation—survival decisions during tumor cell hypoxia. *Mol Cell Biol,* 18, 2845–2854.

Senger, D. R. et al. Tumor cells secrete a vascular permeability factor that promotes accumulation of ascites fluid. *Science* 219, 983 - 985 (1983)

Semenza, G. L., Roth, P. H., Fang, H.-M. & Wang L. W. (1994). Transcriptional regulation of genes encoding glycolytic enzymes by hypoxia-inducible factor 1. *J Biol Chem,* 269, 23757–23763.

Semenza GL, (2003), Targeting HIF-1 for cancer therapy, *Nat. Rev. Cancer*3: 721–732.

Seyfried, T. N., Mukherjee, P., Adams, E., Mulroony, T. & Abate, L. E. (2005) Metabolic Control of Brain Cancer: Role of Glucose and Ketone Bodies. *Proc Amer Assoc Cancer Res,* 46, 1147.

Seyfried, T. N., Sanderson, T. M., El-Abbadi, M. M., McGowan, R. & Mukherjee, P. (2003). Role of glucose and ketone bodies in the metabolic control of experimental brain cancer. *Br J Cancer,* 89, 1375-1382.

Shannon, A. M., Bouchier-Hayes, D. J., Condron, C. M. & Toomey, D. (2003). Tumor hypoxia, chemotherapeutic resistance and hypoxia-related therapies. *Cancer Treat Rev,* 29, 297-307.

Sibold, S., Roh, V., Keogh, A. et al. (2007). Hypoxia increases cytoplasmic expression of NDRG1, but is insufficient for its membrane localization in human hepatocellular carcinoma. *FEBS Lett,* 581, 989–94.

Sobhanifar, S., Aquino-Parsons, C., Stanbridge, E. J. et al. (2005). Reduced expression of hypoxia-inducible factor-1alpha in perinecrotic regions of solid tumors. *Cancer Res,* 65, 7259–66.

Stubbs M, McSheehy PM, Griffiths JR, Bashford CL(2000),Causes and consequences of tumour acidity and implications for treatment, *Mol. Med. Today* 6: 15–19

Supuran CT, Scozzafava A. (2000). Carbonic anhydrase inhibitors: aromatic sulfonamides and disulfonamides act as efficient tumor growth inhibitors. *J Enzyme Inhib.* 15(6): 597 - 610.

Swinson DE, Jones JL, Richardson D, Wykoff C, Turley H, Pastorek J, Harris AL, O'Byrne KJ, (2003), Carbonic anhydrase IX expression, a novel surrogate marker of tumor hypoxia, is associated with a poor prognosis in non-small-cell lung cancer, *J. Clin. Oncol.* 21: 473–482.

StaabA, Löffler J, Said HM,Katzer A, Beyer M, Polat B, Einsele H, Flentje M, Vordermark D, (2007) Modulation of glucose metabolism inhibits hypoxic accumulation of hypoxia-inducible factor-1alpha (HIF-1alpha). *Strahlenther Onkol.* 183(7):366-73.

Stein, S., Thomas, E. K., Herzog, B. et al. (2004). NDRG1 is necessary for p53-dependent apoptosis. *J. Biol. Chem,* 279, 48930–48940.

Stessels F, Van den Eynden G, Van der Auwera I, Salgado R, Van den Heuvel E, Harris AL, Jackson DG, Colpaert CG, van Marck EA, Dirix LY, Vermeulen PB (2004). Breast adenocarcinoma liver metastases, in contrast to colorectal cancer liver metastases, display a non-angiogenic growth pattern that preserves the stroma and lacks hypoxia. *Br. J. Cancer* 90, 1429–1436

Strohman, R. (2002). Maneuvering in the complex path from genotype to phenotype. *Science,* 296, 701-703.

Stryer, L. (1988). *Biochemistry* (3rd edition, pp. 420–1). New York: Freeman and Company.

Svastová, E., Hulíková, A., Rafajová, M., Zatovicová, M., Gibadulinová, A., Casini, A., Cecchi, A., Scozzafava, A., Supuran, C. T., Pastorek, J.,Pastoreková, S. (2004). Hypoxia activates the capacity of tumor-associated carbonic anhydrase IX to acidify extracellular pH, *FEBS Letters,* 577(3), 439–445.

Takeshita, A., Shinjo, K., Naito, K., Ohnishi, K., Higuchi, M. & Ohno, R. (2002). Erythropoietin receptor in myelodysplastic syndrome and leukemia. *Leuk Lymphoma,* 43, 261–264.

Tanimoto, K., Makino, Y., Pereira, T. & Poellinger, L. (2000). Mechanism of regulation of the hypoxia-inducible factor-1 alpha by the von Hippel-Lindau tumor suppressor protein. *EMBO J.,* 19(16), 4298-309.

Toi M, Matsumoto T, Bando H (2001), Vascular endothelial growth factor: its prognostic, predictive, and therapeutic implications. *Lancet Oncol* 2: 667-673.

Teicher BA, Liu SD, Liu JT, Holden SA, Herman TS. (1993), carbonic anhydrase inhibitor as a potential modulator of cancer therapies, *Anticancer Res.* 13:1549 – 1556.

Tisdale, M. J. (1997). Biology of cachexia.. *J Natl Cancer Inst,* 89, 1763-1773.

van Belzen, N. Dinjens, W. N., Diesveld, M. P. et al. (1997). A novel gene which is up-regulated during colon epithelial cell differentiation and down-regulated in colorectal neoplasms. *Lab Invest,* 77, 85–92.

Vaupel, P., Mayer, A. & Hoeckel, M. (2006) Impact of haemoglobin levels on tumor oxygenation: the higher, the better?. *Strahlenther Onkol,* 182, 63–71.

Vaupel, P., Mayer, A. & Hoeckel, M. (2004). Tumor hypoxia and malignant progression. *Methods Enzymol,* 381, 335–354.

Varshney R, Dwarakanath B, Jain V. (2005), Radiosensitization by 6-aminonicotinamide and 2-deoxy D glucose in human cancer cells. *Int J Radiat Biol;* 81: 397 - 408.

Veech, R. L. (2004). The therapeutic implications of ketone bodies: the effects of ketone bodies in pathological conditions: ketosis, ketogenic diet, redox states, insulin resistance, and mitochondrial metabolism. *Prostaglandins Leukot Essent Fatty Acids,* 70, 309-319.

Vordermark, D., Kaffer, A., Riedl, S., Katzer, A. & Flentje, M. (2005). Characterization of carbonic anhydrase IX (CA IX) as an endogenous marker of chronic hypoxia in live human tumor cells. *Int J Radiat Oncol Biol Phys,* 61, 1197–1207.

Vordermark, D., Katzer, A., Baier, K. et al. (2004). Cell-type-specific association of hypoxia- inducible factor-1alpha (HIF-1α) protein accumulation and radiobiologic tumor hypoxia. *Int J Radiat Oncol Biol Phys,* 58, 1242–50.

Vordermark, D., Kraft, P., Katzer, A., Bolling, T., Willner, J. & Flentje, M. (2005). Glucose requirement for hypoxic accumulation of hypoxia-inducible factor-1alpha (HIF-1alpha). *Cancer Lett,* 230, 122–133.

Vordermark, D., Said, H. M., Katzer, A. et. al. (2006). Immunohistochemical detection of osteopontin in advanced head- and-neck cancer: prognostic role and correlation with oxygen electrode measurements, hypoxia-inducible-factor-1alpha-related markers, and hemoglobin levels. *Int J Radiat Oncol Biol Phys.,* 66(5), 1481-7.

Wang, G. L., Jiang, B. H., Rue, E. A. & Semenza, G. L. (1995). Hypoxia-inducible factor 1 is a basic-helix-loop-helix-PAS heterodimer regulated by cellular O2 tension. *PNAS,* 92, 5510-4.

Wang, T., Marquardt, C. & Foker, J. (1976). Nature, 261, 702–705.

Warburg, O. (1925). *Klin Wochenschr Berl,* 4, 534–536.

Warburg O. (1930). *The metabolism of tumors,* Constable Press.

Westenfelder, C., Baranowski, R. L. (2000). Erythropoietin stimulates proliferation of human renal carcinoma cells. *Kidney Int,* 58, 647–657.

Welsh S, Williams R, Kirkpatrick L, Paine-Murrieta G, Powis G.(2004),Antitumor activity and pharmacodynamic properties of PX - 478, an inhibitor of hypoxia-inducible factor-1alpha. *Mol Cancer Ther,*3: 233 - 244.

Wingo, T., Tu, C., Laipis, P. J., Silverman, D. N. (2001). The catalytic properties of human carbonic anhydrase IX. *Biochem Biophys. Res. Commun.* 288, 666-669.

Wykoff, CC., Beasley, N., Watson, P., Turner, L., Pastorek, J., Wilson, G., Turley, H., Maxwell, P., Pugh, C., Ratcliffe, P. & Harris, A. (2000). Hypoxia-inducible regulation of tumor-associated carbonic anhydrases. *Cancer Res.,* 60, 7075-7083.

Yasuda, Y., Fujita, Y., Masuda, S., Musha, T., Ueda, K., Tanaka, H., Fujita, H., Matsuo, T., Nagao, M., Sasaki, R. & Nakamura, Y. (2002). Erythropoietin is involved in growth and angiogenesis in malignant tumours of female reproductive organs. *Carcinogenesis,* 23, 1797–1805.

Yasuda, Y., Musha, T., Tanaka, H., Fujita, Y., Fujita, H., Utsumi, H., Matsuo, T., Masuda, S., Nagao, M., Sasaki, R. & Nakamura, Y.(2001). Inhibition of erythropoietin signalling destroys xenografts of ovarian and uterine cancers in nude mice. *Br J Cancer,* 84, 836–843.

Yasuda S, Arii S, Mori A, et al. (2004), Hexokinase II and VEGF expression in liver tumors: correlation with hypoxia-inducible factor-1a and its significance.*J Hepatol;* 40: 117–23.

Yun H, Lee M, Kim SS, et al. (2005), Glucose deprivation increases mRNA stability of vascular endothelial growth factor through activation of AMP-activated protein kinase in DU145 prostate carcinoma. *J Biol Chem;* 280: 9963 -72

Závada, J., Závadová, Z., Pastorek, J., Biesová, Z., Jezek, K. & Velek, J. (2000). Human tumour-associated cell adhesion protein mediating cell adhesion. *Br. J Cancer,* 82, 1808-1813.

Závada, J., Závadová, Z., Pastoreková, S., Ciampor, F., Pastorek, J. & Zelník, V. (1993). Expression of MaTu-MN protein in human tumor cultures and in clinical specimens. *Int. J. Cancer,* 54, 268-274.

Zhang, P., Tchou-Wong, K. M. & Costa, M. (2007). Egr-1 mediates hypoxia-inducible transcription of the NDRG1 gene through an overlapping Egr-1/Sp1 binding site in the promoter. *Cancer Res,* 67(19), 9125-9133.

Index

A

acid, x, 7
acidity, 43
acidosis, 2, 5, 9
active transport, 6
acute lymphoblastic leukemia, 18
adaptation, ix, 22, 27, 32
adenocarcinoma, 44
adenosine, 31
adhesion, xi, 46
aggressiveness, 13
anemia, 18
angiogenesis, x, 14, 18, 25, 32, 33, 34, 35, 37, 38, 41, 46
anhydrase, x, xi, 3, 6, 12, 15, 20, 21, 33, 36, 38, 40, 41, 42, 43, 44, 45
antagonism, 3
anticancer drug, xii, 2, 6, 7, 27
antigen, 33
apoptosis, 2, 5, 9, 11, 22, 35, 40, 44
arrest, 41
arteriovenous shunt, ix, 22
aryl hydrocarbon receptor, ix
ascites, xii, 14, 43
astrocytoma, vii, xii, 20, 21, 22, 39
ATP, xii, 1, 2, 5, 7, 8, 9, 13, 40, 41

B

background, 12, 27
bacteriostatic, 3
behavior, 8
benzodiazepine, 2
bicarbonate, viii, xi, 6
binding, vii, ix, 18, 25, 26, 33, 34, 38, 40, 41, 46
biological processes, x, 17
biosynthesis, 6
bladder, xi
blocks, vii, x, xi
blood, ix, 2, 14, 22, 40
blood flow, 22, 40
blood supply, 14
bone, 14
bone marrow, 14
brain, viii, xi, 1, 2, 5, 7, 9, 18, 21, 23, 27, 33, 38, 40, 43
brain tumor, viii, 5, 7, 9, 18, 27, 38
breast cancer, 14, 32, 33, 39
breast carcinoma, 18, 31, 32

C

cachexia, 44
calcium, 31
cancer cells, xii, xiii, 2, 4, 11, 13, 15, 22, 31, 41, 44

48　　　　　　　　　　　　　　　　　Index

capillary, 2
carbon, x, xii, 1, 6
carbon dioxide, x, 6
carcinogenesis, 31, 39, 42
carcinoma, 18, 33, 45
catalytic activity, 5, 9, 13
catalytic properties, 45
cell cycle, 37, 39
cell line, 3, 5, 9, 12, 16, 19, 27, 42
cell lines, 3, 5, 9, 12, 16, 19, 27, 42
cell metabolism, viii, 27
cell surface, xi
cellular regulation, 39
cervical cancer, 31, 40
cervix, xi, 18, 31, 36, 37
chemical stability, 12
chemotherapeutic agent, 3
chemotherapy, 4, 14, 32
chloroform, 33
choroid, xi
chronic hypoxia, 45
chronic renal failure, 18
clinical trials, 15
CO2, 7
colon, 44
colorectal cancer, 15, 44
complementary DNA, 41
concentration, 3, 5, 9, 19
consumption, xii, 1
control, 2, 8, 16, 19, 20, 21, 23, 37, 43
control group, 23
correlation, vii, ix, 8, 20, 21, 22, 32, 40, 45, 46
cycles, 13
cytochrome, 2
cytotoxicity, 4, 31, 39

D

decisions, 43
defects, 2
degradation, vii, x, xiii, 7, 25
degradation process, x
density, 33
deposition, 33
deposits, 33
depression, 40

deprivation, vii, ix, 4, 31, 46
derivatives, 11
detection, 32, 42, 45
diagnostic markers, 17
diet, 45
differentiation, 17, 22, 31, 38, 41, 44
diffusion, ix, 2, 6, 22
disease progression, xiii, 3
distribution, xi
DNA, 41
dominance, 37
drug resistance, xii
drugs, 3, 6, 7, 11
duration, 18
dysplasia, 31

E

electrodes, 42
emission, 1
encoding, ix, 22, 43
endothelial cells, 2, 14
environment, vii, ix, 6, 8, 12, 13, 15
enzymatic activity, xi, 7
enzyme inhibitors, xii
enzymes, 5, 9, 13, 36, 37, 43
epilepsy, 35
epithelia, xi
epithelial cells, xi
erythrocytosis, 18
erythropoietin, 20, 21, 31, 32, 34, 36, 39, 40, 46
evolution, 41
extraction, 1, 33

F

failure, 15, 38
family, x, 14, 22, 40, 41
fibrin, 33
fibroblast growth factor, 38
fibrosarcoma, 5
flexibility, 7
fluid, 43
France, 38
free radicals, 39

G

gallbladder, xi
gene expression, 4, 16, 17, 19, 25, 33, 34, 35, 41, 42
generation, 7
genotype, 44
glia, 2
glioblastoma, viii, xii, xiii, 3, 9, 17, 20, 21, 23, 26, 42
glioma, viii, 5, 9, 14, 16, 27, 40, 42
gluconeogenesis, 6
glucose, viii, x, xi, xii, 1, 2, 3, 4, 5, 7, 9, 31, 38, 39, 40, 41, 42, 43, 44
GLUT, 2, 39
glutamate, 42
glycerol, 42
glycolysis, viii, xii, 1, 2, 3, 4, 5, 7, 8, 9, 13, 16, 27, 33, 35, 39, 40
glycosylation, 34
grading, 20, 21
groups, 15, 18, 22, 26, 29
growth, vii, viii, ix, xii, 6, 14, 18, 19, 22, 25, 27, 31, 33, 34, 35, 38, 39, 40, 41, 44, 46
growth factor, xii, 14, 18, 19, 35, 38, 39, 41, 44, 46
growth rate, 14
guidance, 34

H

head and neck cancer, 38
hemoglobin, 32, 40, 45
hepatocellular carcinoma, 43
hormone, 17
host, 22
human brain, vii, xi, xii, 16, 17, 19, 20, 21, 27, 42
hypoxia, vii, viii, ix, x, xi, xii, xiii, 3, 4, 5, 6, 7, 9, 11, 12, 13, 14, 16, 17, 18, 19, 20, 21, 22, 25, 26, 27, 29, 31, 32, 33, 34, 35, 36, 37, 38, 39, 40, 41, 42, 43, 44, 45, 46
hypoxia-inducible factor, x, xiii, 14, 20, 21, 31, 32, 35, 37, 38, 39, 40, 43, 44, 45, 46
hypoxic cells, 2, 5, 40

I

identification, 15
immunohistochemistry, 18
in vitro, viii, xi, 5, 6, 8, 9, 15, 16, 17, 18, 19, 26, 32, 35, 41, 42
in vivo, viii, xii, 6, 8, 17, 20, 23, 35, 41, 42
inducer, 38
inducible protein, vii, x, 31
induction, viii, xi, 4, 8, 21
inefficiency, 7
inhibition, viii, xii, xiii, 2, 3, 4, 5, 6, 7, 27, 34, 39
inhibitor, 3, 5, 9, 15, 16, 38, 41, 44, 45
initiation, 15
injury, iv, 33
insulin, 45
insulin resistance, 45
interaction, xi, 4
interactions, 8
interference, xii, 3
ions, 7
iron, x
isolation, 33
isozyme, 41
isozymes, 7

K

kidney, xi
Krebs cycle, 2

L

lactate dehydrogenase, 5, 9, 13
lactic acid, 5, 7, 9, 13, 22
leukemia, 18, 33, 44
links, 17
lipids, vii, 1
liver, 15, 44, 46
liver metastases, 44
localization, 43
lung cancer, 38, 43
lymph, 32
lymph node, 32
lysine, x

M

machinery, 12
macromolecules, vii, 1
magnetic resonance, 35
maintenance, 4
malignancy, 8
management, 6, 7, 8, 35
manipulation, viii, 5
marker genes, vii
matrix, 34
metabolic pathways, 1, 8, 13
metabolism, x, xiii, 1, 2, 3, 5, 6, 7, 9, 13, 33, 35, 39, 40, 41, 43, 45
metabolites, 4, 42
metastasis, ix, 14, 18, 27, 33, 34, 42
mice, 18, 39, 46
microdialysis, 42
microenvironments, xi
mitochondria, 2, 39
mitogen, viii, xii, 14
model, 15, 25
models, 5, 6, 7, 15
mole, 7, 8
molecules, xiii, 14
monoclonal antibody, 33
morphogenesis, 34
mortality, 15
mRNA, viii, xii, 3, 4, 5, 9, 13, 16, 17, 19, 20, 21, 23, 39, 46

N

neck cancer, 32, 45
necrosis, 33
neovascularization, 14
neurons, 2
nickel, 42
nitric oxide, 36, 39, 40
nitric oxide synthase, 40
nucleotides, vii, 1
nucleus, vii, x
nutrients, vii, ix, 22

O

ovarian tumor, 32
oxidative stress, 4, 31
oxygen, vii, ix, x, xi, xii, xiii, 1, 4, 5, 7, 8, 9, 11, 13, 22, 26, 32, 36, 39, 40, 42, 45
oxygen consumption, 4

P

p53, 22, 44
pancreatic cancer, 31, 38
parameter, 31
parenchymal cell, 15
passive, 6
pathways, x, 4, 5, 8, 9, 13, 34, 38, 39, 41
PCR, 4, 19
permeability, 14, 34, 43
pH, xii, 5, 6, 7, 9, 11, 13, 44
phenol, 33
phenotype, viii, xii, 8, 15, 17, 44
phosphorylation, x, 4, 40
plasma, 38, 42
platinum, 3
plexus, xi
PM, 35, 43
polypeptide, x
poor, ix, xii, 1, 6, 27, 31, 43
prediction, 13
primary brain tumor, 1
production, xii, 2, 5, 7, 8, 9, 13, 22
prognosis, vii, ix, xiii, 1, 14, 22, 27, 31, 32, 33, 37, 43
proliferation, vii, ix, x, xii, 1, 6, 8, 14, 17, 18, 39, 43, 45
promoter, vii, x, 25, 40, 46
propagation, 13
prostate, 32, 46
prostate cancer, 32
prostate carcinoma, 46
protein synthesis, 4, 40
proteins, vii, ix, x, xi, 1, 5, 9, 22, 25, 26, 39
protons, 6
proto-oncogene, 36

Index

R

radiation, xii, 2, 4, 15, 35, 40, 42
Radiation, 29, 39
radiation therapy, xii, 2, 40, 42
radio, xiii, 6, 15
radiosensitization, 39
radiotherapy, xii, 5, 14, 31, 32, 35, 38, 39
receptors, 2
recognition, 25
region, vii, x, xi, 25
regulation, xi, xii, 3, 4, 12, 13, 14, 19, 22, 25, 26, 29, 35, 37, 41, 42, 43, 44, 45
regulators, 4, 38
relationship, 33, 40
remission, 18
repair, 37
reproductive organs, 46
residues, ix
resistance, xii, 2, 6, 11, 13, 14, 36, 43
respiration, 3, 8
respiratory, 7

S

selectivity, xii, 7
sensing, 32
sensitivity, xiii
sensors, 35
side effects, 7
signal transduction, 4
signalling, 15, 29, 35, 46
signals, 35
siRNA, xiii, 27
small intestine, xi
smooth muscle, 35
smooth muscle cells, 35
solid tumors, 5, 14, 22, 43
solubility, 12
squamous cell, 31, 33, 34
squamous cell carcinoma, 31, 33, 34
stability, x, xi, 13, 38, 46
stabilization, 3
stasis, 15
stomach, xi, 41
strategies, 1, 17, 26, 32
stress, vii, ix, xii, 4, 22, 27, 39
stress granules, 39
stroma, 44
sulfonamide, 7
sulfonamides, 3, 7, 43
Sun, 37
supply, vii, ix, 7, 22
survival, vii, viii, ix, x, xii, xiii, 11, 13, 14, 17, 32, 33, 35, 36, 39, 43
synergistic effect, 6
synthesis, vii, 1

T

targets, x, 17, 25, 26, 27
technical assistance, 29
tension, xiii, 45
therapeutic approaches, viii, xii, 3, 6, 27
therapeutic targets, vii
therapy, viii, xii, 11, 14, 18, 27, 33, 41, 43
tissue, vii, viii, ix, xi, 1, 2, 14, 17, 18, 22, 23, 32
toxicity, 12
transcription, vii, ix, xi, xiii, 6, 11, 12, 22, 25, 26, 37, 39, 46
transcription factors, 22
transcripts, 18
transformation, 1, 2, 5, 9, 13, 18, 33, 39
transition, 8
translation, 4, 39
translocation, vii, x
transport, 40
trial, 38
tricarboxylic acid, 7
triggers, 2, 5, 9, 25
tumor cells, viii, xi, xii, 1, 5, 6, 7, 8, 9, 15, 19, 22, 27, 32, 34, 39, 42, 45
tumor growth, x, 3, 4, 5, 15, 18, 43
tumor metastasis, 32
tumor progression, viii, xii, 11, 22, 31
tumor resistance, 35
tumorigenesis, 6, 7, 11
tyrosine, 15

U

ultrastructure, 40
uterine cancer, 46

V

vascular endothelial growth factor (VEGF), 14, 20, 21, 39
vasculature, ix, 15, 22
vasopressor, 36
VEGF expression, 14, 21, 46
vessels, viii, xii, 14, 36
vestibular schwannoma, 18, 34
vitamin D, 41

X

xenografts, 46